Introduction to Computer Organization

FIRST EDITION

Gang Hu, Guanqiu Qi, Neal Mazur and Zhiqin Zhu

Buffalo State University

cognella

SAN DIEGO

Bassim Hamadeh, CEO and Publisher
John Remington, Managing Executive Editor
Kaela Martin, Project Editor
Susana Christie, Senior Developmental Editor
Abbey Hastings, Production Editor
Emely Villavicencio, Senior Graphic Designer
JoHannah McDonald, Licensing Coordinator
Natalie Piccotti, Director of Marketing
Kassie Graves, Senior Vice President, Editorial

Printed in the United States of America.

cognella® | ACADEMIC PUBLISHING
320 South Cedros Ave., Ste. 400, Solana Beach, CA 92075

Contents

CHAPTER

1

Computer Numbering Systems

Introduction

Computers were initially developed to process numerical data with greater speed and accuracy than could be accomplished by humans. The first step in this endeavor was to develop methods to represent the data efficiently and with acceptable precision. The techniques for accomplishing these goals caused computer designers to think about numerical data representation in different ways, given restrictions of digital technology and available space. Many problems arose concerning the representation of different types of numbers that are needed in computation, such as integers, negative numbers, and fractional values. Solutions included changing number bases and the use of scientific notation representation. Concerns associated with the efficient processing of the numbers further influenced the representations that were developed. As computers began to be used for applications other than mathematical computations, other data, such as text, needed to be given standard representations. The foundations of data representation need to be understood before moving on to other computer organization topics such as digital logic, memory, processing units, and data communication within the system.

Objectives

This chapter will help students:

- **understand** the fundamentals of numerical data representation and manipulation in digital computers.

- **master** the skill of converting between decimal and binary number systems.
- **gain** familiarity with some common binary numbers, such as numbers with decimal values of 0 to 15, 127, 255, etc.
- **understand** the motivation of utilizing octal and hexadecimal number systems.
- **master** the skill of converting among hexadecimal, octal, and binary numbers.
- **grasp** basic binary arithmetic calculations.

Key Terms

positional weighting: the value of a digit in a sequence is determined by the digit number and its position associated with a weight within the sequence. Commonly, the value is digit number multiplied by the weight.

binary number: number expressed in the base-2 numeral system, which uses only two symbols: 0 and 1.

octal number: number expressed in the base-8 numeral system, which uses eight symbols: 0, 1, 2, 3, 4, 5, 6, 7.

hexadecimal: number expressed in the base-16 numeral system, which uses 16 symbols: 0 to 9 and A to F.

Positional Numbering Systems

Arabic numbers consist of 10 digits: 0, 1, 2, 3, 4, 5, 6, 7, 8, and 9. These 10 digits are used in the decimal number system. The decimal Hindu-Arabic number system was developed in India around 700 AD. It was modified into Arabic numerals later in North Africa. Europe started to use the decimal Hindu-Arabic number system in the middle of the 16th century.

All of the number systems that will be presented use **positional weighting**. For example, each digit in the decimal number 1,677 has a weight associated with it. There is a 1 in the thousands position (1000), 6 in the hundreds position (600), 7 in the tens position (70), and 7 in the ones position (7). These values are added to give the full value of the number. Although both the tens digit and ones digit have the same number, 7, the value of the tens digit (70) is ten times that of the ones digit (7).

Positional numbering systems are proposed to present any numeric value through increasing powers of a base (or radix). Each position of a numeric value is weighted by a power of the corresponding base.

The base size of a positional numbering system determines the set of valid numerals. For example, there are two digits for the binary (base 2) number system, 0 and 1. A single digit in the binary number system is often referred to as a bit. There are eight digits for the octal (base 8) number system, 0 to 7. There are 10 digits in the decimal (base 10) number system, 0 to 9. There are 16 digits in the hexadecimal (base 16) number system, 0 to 9, and A, B, C, D, E, and F represent the decimal 10, 11, 12, 13, 14, and 15, respectively. Binary, octal, and hexadecimal number systems are three commonly used number systems in computer science. Computers only use the binary number system to manipulate and store all data. This includes numbers, words, videos, graphics, and music. The choice of the binary number system is predicated on the fact that it is easy to build a machine that represents only two states. Think of a light switch. **Binary numbers** tend to be extremely long even for small values, and thus octal and hexadecimal representations are often used to provide an easily derived compact representation of binary-coded values.

Since 0 is the initial valid number, the largest valid digit in a positional numbering system is 1 less than the base size. For example, 8 is not a valid number in any positional numbering system with the base size less than 9. The base is used as a subscript to present the corresponding positional numbering system. For example, 62_8 represents the **octal number** 62. Similarly, 83_{10} represents the decimal number 83. The decimal number system is the default for humans, so the subscript 10 is generally not included. This book will follow this convention. Any decimal integer can be expressed exactly in any other integral base system by the corresponding conversion (see Example 1.1).

Example 1.1: Five numbers represented as powers of a base.

$$372.49_{10} = 3 \times 10^2 + 7 \times 10^1 + 2 \times 10^0 + 4 \times 10^{-1} + 9 \times 10^{-2}$$

$$10011_2 = 1 \times 2^4 + 0 \times 2^3 + 0 \times 2^2 + 1 \times 2^1 + 1 \times 2^0 = 19_{10}$$

$$11101.01_2 = 1 \times 2^4 + 1 \times 2^3 + 1 \times 2^2 + 0 \times 2^1 + 1 \times 2^0 + 0 \times 2^{-1} + 1 \times 2^{-2} = 29.25_{10}$$

$$2347_8 = 2 \times 8^3 + 3 \times 8^2 + 4 \times 8^1 + 7 \times 8^0 = 1255_{10}$$

$$3B9_{16} = 3 \times 16^2 + 11 \times 16^1 + 9 \times 16^0 = 953_{10}$$

Binary Numbering System

Although the first electronic computer, ENIAC (short for *Electronic Numerical Integrator and Calculator*), was created in 1946 at the University of Pennsylvania, the binary numbering system was invented almost three centuries earlier. The binary numbering system was first published in the paper *Essay d'une nouvelle science des nombres* authored by Gottfried Wilhelm Leibniz (1646–1716) (the coinventor of *calculus*). However, the actual discovery of the binary numbering system occurred more than 20 years earlier.

TABLE 1.1 Some Numbers to Remember

POWERS OF 2
$2^{-2} = 1/4 = 0.25$
$2^{-1} = 1/2 = 0.5$
$2^{0} = 1$
$2^{1} = 2$
$2^{2} = 4$
$2^{3} = 8$
$2^{4} = 16$
$2^{5} = 32$
$2^{6} = 64$
$2^{7} = 128$
$2^{8} = 256$
$2^{9} = 512$
$2^{10} = 1024$
...
$2^{16} = 65{,}536$

DECIMAL	4-BIT BINARY	HEXADECIMAL
0	0000	0
1	0001	1
2	0010	2
3	0011	3
4	0100	4
5	0101	5
6	0110	6
7	0111	7
8	1000	8
9	1001	9
10	1010	A
11	1011	B
12	1100	C
12	1101	D
14	1110	E
15	1111	F

The binary numbering system is the simplest positional numbering system. Its base is 2. There are only two digits (0 and 1) used in the binary representation of any number. Because of its simplicity, the binary numbering system can be easily applied to electronic circuitry. So, the binary system builds the foundation of modern electronic digital computers. The logic gates used in microchips only have two states (off and on) that are associated with digits 0 and 1. The state of a logic gate can be represented by one bit (a binary digit). Along with on/off, a digit in the binary system can be used to represent any variable with two values such as yes/no. One of the most important ways of interpreting binary 1/0 is representing the logical values true/false.

After some practice, the binary numbering system can be understood by humans. Small binary numbers (shown in Table 1.1) can be easily recognized at a glance. The conversion of large values with more digits is more complex. Octal and hexadecimal numbering systems are used as compressed representations of binary numbers. The conversion between two different positional numbering systems is discussed in the following sections.

Binary (Base 2)

This section only discusses the base conversion of unsigned numbers. Representation and conversion of negative numbers will be shown later.

Binary to Decimal Conversion

Similar to **Example 1.1**, addition can be applied to convert numbers from binary to decimal. The decimal value is obtained by adding the value of each unit digit of a binary number. Since we are working in base 2, powers of 2 are used in computing weight values. The computation of binary values is simplified because binary digits only consist of values of 1 (add digit weight to result) or 0 (don't add the weight to the result).

Example 1.2: Convert 11010011_2 to base 10.

$$11010011_2 = 1 \times 2^7 + 1 \times 2^6 + 0 \times 2^5 + 1 \times 2^4 + 0 \times 2^3 + 0 \times 2^2 + 1 \times 2^1 + 1 \times 2^0$$

$$= 128 + 64 + 0 + 16 + 0 + 0 + 2 + 1$$

$$= 211_{10}$$

Decimal to Binary Conversion

The division-remainder method is effective in converting a number between different base systems. The corresponding details are discussed in examples that follow.

The division-remainder method can be used to convert a decimal number to a binary number. There is no need to know the powers of the base. Successive divisions by the base are used giving a quotient and remainder at each step as shown in Figure 1.1. For binary numbers, note that odd numbers have a remainder of 1 and even numbers have a remainder of 0. Quotients are obtained through an integer division. Decimal division results are thus rounded down. The process of dividing resulting quotients continues until the quotient reaches 0. After the successive divisions, the remainders from bottom to top give the conversion result.

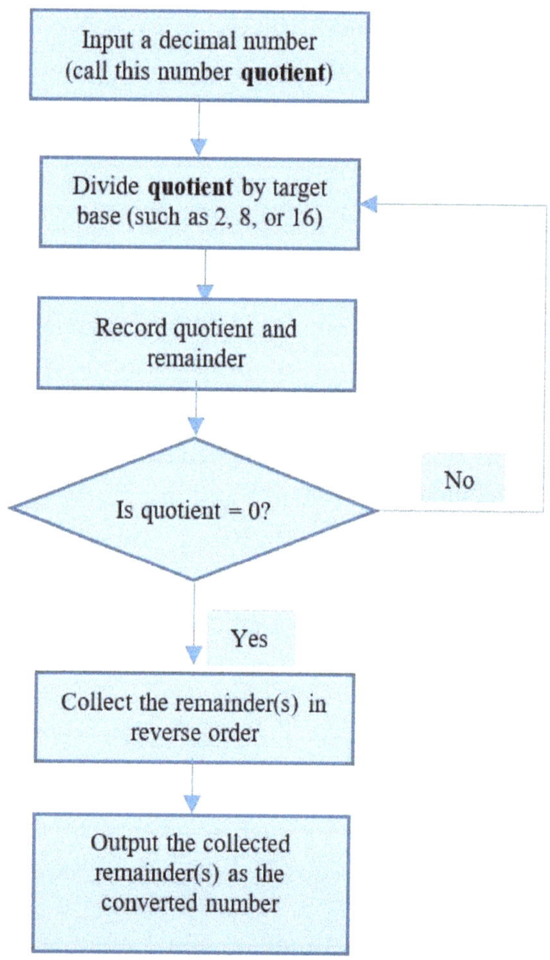

FIGURE 1.1 Flowchart of Division-Remainder Algorithm for Converting Decimal to Other Bases

For example, let's convert 125_{10} to base 2. As shown in Example 1.3, the details of the conversion from 125_{10} to base 2 are processed by the division-remainder method.

Example 1.3: Convert 125_{10} to base 2 using the division-remainder method.

2 ⌊ 125	...1	125 / 2 = 62 with a remainder of 1
2 ⌊ 62	...0	62 / 2 = 31 with a remainder of 0
2 ⌊ 31	...1	31 / 2 = 15 with a remainder of 1
2 ⌊ 15	...1	15 / 2 = 7 with a remainder of 1
2 ⌊ 7	...1	7 / 2 = 3 with a remainder of 1
2 ⌊ 3	...1	3 / 2 = 1 with a remainder of 1
2 ⌊ 1	...1	1 / 2 = 0 with a remainder of 1
0		

Reading the remainders from bottom to top, we can get $125_{10} = 1111101_2$. Convert this binary result back to decimal to convince yourself that the algorithm worked.

This method works with any base. Example 1.4 shows another decimal to binary conversion example.

Example 1.4: Convert 159_{10} to base 2 using the division-remainder method.

2 ⌊ 159	...1	159 / 2 = 79 with a remainder of 1
2 ⌊ 79	...1	79 / 2 = 39 with a remainder of 1
2 ⌊ 39	...1	39 / 2 = 19 with a remainder of 1
2 ⌊ 19	...1	19 / 2 = 9 with a remainder of 1
2 ⌊ 9	...1	9 / 2 = 4 with a remainder of 1
2 ⌊ 4	...0	4 / 2 = 2 with a remainder of 0
2 ⌊ 2	...0	2 / 2 = 1 with a remainder of 0
2 ⌊ 1	...1	1 / 2 = 0 with a remainder of 1
0		

Reading the remainders from bottom to top, we can get $159_{10} = 10011111_2$.

Fractional Binary to Decimal Conversion

Addition can also be applied to the conversion from fractional binary to fractional decimal. The fractional decimal value is obtained by adding the value of each unit digit of a fractional binary number. Weights on fractional binary values are represented by negative powers applied to the base 2. Table 1.2 shows the basic conversion process from fractional binary to decimal.

TABLE 1.2 Conversion Fractional Binary to Decimal

FRACTIONAL BINARY	CONVERSION	FRACTIONAL DECIMAL
0_2	0×2^0	0_{10}
0.1_2	1×2^{-1}	0.5_{10}
0.01_2	1×2^{-2}	0.25_{10}
0.001_2	1×2^{-3}	0.125_{10}
0.0001_2	1×2^{-4}	0.0625_{10}
0.00001_2	1×2^{-5}	0.03125_{10}
0.000001_2	1×2^{-6}	0.015625_{10}
0.0000001_2	1×2^{-7}	0.0078125_{10}
0.00000001_2	1×2^{-8}	0.00390625_{10}
0.000000001_2	1×2^{-9}	0.00195312_{10}

Example 1.5: Convert 0.101_2 to base 10.

$$0.101_2 = 1 \times 2^{-1} + 0 \times 2^{-2} + 1 \times 2^{-3} = 1 \times 0.5 + 0 + 1 \times 0.125 = 0.625_{10}$$

Fractional Decimal to Binary Conversion

To translate from a fractional decimal number to a binary number, the decimal value is repeatedly multiplied by 2. The integer part of each product, 0 or 1, is included in the binary result. These integer parts in the order generated are placed after the decimal point to derive the binary fractional result. The process stops when either the fractional part of the multiplication result is 0 or the desired number of binary places is reached.

Example 1.6: Convert 0.125_{10} to base 2.

$$
\begin{array}{r l}
0.125 & \\
\times \quad 2 & \\
\hline
0.25 & \text{integer part: 0, fractional part: 0.25} \\
\times \quad 2 & \\
\hline
0.5 & \text{integer part: 0, fractional part: 0.5} \\
\times \quad 2 & \\
\hline
1.0 & \text{integer part: 1, fractional part: 0}
\end{array}
$$

Reading the value of the integer part from top to bottom, we get $0.125_{10} = 0.001_2$.

Example 1.7: Convert 0.22_{10} to base 2.

$$
\begin{array}{r l}
0.22 & \\
\times \quad 2 & \\
\hline
0.44 & \text{integer part: 0, fractional part: 0.44} \\
\times \quad 2 & \\
\hline
0.88 & \text{integer part: 0, fractional part: 0.88} \\
\times \quad 2 & \\
\hline
1.76 & \text{integer part: 1, fractional part: 0.76} \\
0.76 & \text{only the fractional part is used to do the multiplication} \\
\times \quad 2 & \\
\hline
1.52 & \text{integer part: 1, fractional part: 0.52}
\end{array}
$$

In **Example 1.7,** 0.22_{10} is different from the previous example, 0.125_{10}. The fractional part of the repeated multiplication results of 0.22_{10} will not converge to 0. In this case, non-convergence after repeated multiplications, the fractional decimal number does not translate exactly into binary. So, users can only get an approximate value according to their need for accuracy.

Reading the value of the integer part from top to bottom, we get $0.1875_{10} = 0.0011_2$, which approximates in four digits the decimal fraction (stopping after four rounds of repeated multiplications). The more bits that are used, the more accurate the result will be. There is a round-off error between the actual decimal number and the converted decimal number from the approximated binary number.

$$0.0011_2 = 0 \times 2^{-1} + 0 \times 2^{-2} + 1 \times 2^{-3} + 1 \times 2^{-4} = 0 + 0 + 1 \times 0.125 + 1 \times 0.0625 = 0.1875_{10}$$

$$\begin{array}{r} 0.2200_{10} \\ -\ 0.1875_{10} \\ \hline 0.0325_{10} \end{array} \quad \text{(round-off error)}$$

In this **Example 1.7**, 0.0325_{10} is the round-off error.

Example 1.8: Convert 3.14_{10} to base 2.

3.14_{10} consists of both integer and fractional parts. According to the above discussion, different methods are used for converting integer and fractional parts.

integer part:

$$\begin{array}{l} 2 \,\underline{|\ 3\quad}\ \ldots 1 \quad 3/2 = 1 \text{ with a remainder of } 1 \\ \quad 2 \,\underline{|\ 1\quad}\ \ldots 1 \quad 1/2 = 0 \text{ with a remainder of } 1 \\ \qquad 0 \end{array}$$

The integer part $3_{10} = 11_2$.

fractional part:

$$\begin{array}{rl} 0.14 & \\ \times\ \ 2 & \\ \hline 0.28 & \text{integer part: 0, fractional part: 0.28} \\ \times\ \ 2 & \\ \hline 0.56 & \text{integer part: 0, fractional part: 0.56} \\ \times\ \ 2 & \\ \hline 1.12 & \text{integer part: 1, fractional part: 0.12} \\ 0.12 & \text{only the fractional part is used to do the multiplication} \\ \times\ \ 2 & \\ \hline 0.24 & \text{integer part: 0, fractional part: 0.52} \end{array}$$

According to the above discussion, the four digits in the fractional part are only an approximation (stopping after four rounds of repeated multiplications). So, 0.14 is approximated by $0.0010_2 = 0.125_{10}$. There will be a round-off error of 0.14 – 0.125 = 0.015.

After combining the integer and fractional parts of the converted binary numbers, $3.14_{10} = 11.0010_2$.

Octal (Base 8)

Octal numbers were widely used in computer systems/machines involving 6-bit, 12-bit, 24-bit, or 36-bit words such as the UNIVAC 1050, PDP-8, ICL 1900, and IBM mainframes. Binary numbers are too complex to use. Decimal numbers need to be converted by complex hardware. Hexadecimal numbers involve more numerals. Since the above word size is divisible by 3 (each octal digit consists of three binary digits), octal numbers are good to shorten the word size of these machines. Three bits are used to represent any octal digit in binary (from 000 to 111). Conversions from binary to octal, and vice versa, are easy to implement.

Octal to Decimal Conversion

Like binary conversion to decimal, addition is used when converting from octal to decimal. The decimal value is obtained by adding the products of the octal digit times the value of each unit digit of an octal number. In this case, the unit digit is computed by raising the base 8 to the power of its positional value.

Example 1.9: Convert 163_8 to base 10.

$$163_8 = 1 \times 8^2 + 6 \times 8^1 + 3 \times 8^0$$

$$= 1 \times 64 + 6 \times 8 + 3 \times 1 = 115_{10}$$

Decimal to Octal Conversion

The division-remainder method can be applied to the conversion from decimal to octal. The decimal number is successively divided by the base 8 until the remainder reaches 0. After the successive divisions, the remainders from bottom to top give the result of the conversion.

Example 1.10: Convert 125_{10} to base 8 using the division-remainder method.

8 | 125	… 5	125 / 8 = 15 with a remainder of 5
8 | 15	… 7	15 / 8 = 1 with a remainder of 7
8 | 1	… 1	7 / 8 = 0 with a remainder of 1
0		

Reading the remainders from bottom to top, we can get $125_{10} = 175_8$.

Binary to Octal Conversion

Groups of three digits of a binary number can represent a single base 8 digit from 000 (0 in base 8) to 111 (7 in base 8). A binary number is divided into one three-digit binary number or multiple three-digit binary numbers starting from the rightmost digit to the leftmost digit. If the remaining number of digits is less than three in the division process, 0(s) is added to the leftmost of the remaining digit(s) to compose the three digits.

Example 1.11: Convert 10110101_2 to base 8.

Each group of three digits of 10110101_2 from right to left is converted to the corresponding unit digit of base 8. The leftmost two digits are 10. So, 0 is added to the leftmost 10_2 to create 010_2, which has three digits.

From the right to left 10110101_2,

$$10\ 110\ 101_2$$

0 is added to the leftmost of 10, which are the leftmost two digits.

$$\Rightarrow 010\ 110\ 101_2$$

$$2_8\ \ 6_8\ \ 5_8$$

$$101_2 = 1 \times 2^2 + 0 \times 2^1 + 1 \times 2^0 = 5,$$

$$110_2 = 1 \times 2^2 + 1 \times 2^1 + 0 \times 2^0 = 6,$$

010_2 (0 marked in red is added to complement) $= 0 \times 2^2 + 1 \times 2^1 + 1 \times 2^0 = 2.$

The result is: $10110101_2 = 265_8$.

Octal to Binary Conversion

Each digit of an octal number can be represented as a three-digit binary number. If the number of digits of a binary number is less than three, 0(s) should be added to the leftmost position(s) of the binary number. Only 0(s) at the leftmost position(s) of the converted binary number can be waived.

Example 1.12: Convert 364_8 to base 2.

$$364_8$$

$$=> 011_2\ 110_2\ 100_2$$

364_8 –> three digits are processed separately, 3_8–> 011_2 (0 marked in red shown in the highest position of the converted binary number can be waived), 6_8–> 110_2, 4_8–> 100_2

So, 364_8=11110100_2.

Hexadecimal (Base 16)

Generally, the hexadecimal numbering system is used to describe locations in computer memory. Hexadecimal represents an even more compact representation of binary numbers. Remember that there are 16 digits in the hexadecimal numbering system, which are the digits 0 to 9 followed by digits A, B, C, D, E, and F.

Hexadecimal to Decimal Conversion

Like binary conversion to decimal, addition is used when converting from hexadecimal to decimal. The decimal value is obtained by adding the products of the hexadecimal digit times the value of each unit digit of a **hexadecimal number**. In this case, the unit digit is computed by raising the base 16 to the power of its positional value.

Example 1.13: Convert $2B5_{16}$ to base 10.

$$2B5_{16} = 2 \times 16^2 + 11 \times 16^1 + 5 \times 16^0 = 512 + 176 + 5 = 693$$

Decimal to Hexadecimal Conversion

The division-remainder method can be applied to the conversion from decimal to hexadecimal. The decimal number is successively divided by the base 16 until the remainder reaches 0. After the successive divisions, the remainders from bottom to top give the result of the conversion.

Example 1.14: Convert 252_{10} to base 16 using the division-remainder method.

16 | 252	… 12	252/ 16 = 15 with a remainder of 12
16 | 15	… 15	15 / 16 = 0 with a remainder of 15
0		

Reading the remainders from bottom to top, we can get $FC_{16} = 15 \times 16^1 + 12 \times 16^0 = 240 + 12 = 252_{10}$.

Binary to Hexadecimal Conversion

Similar to the conversion from binary to octal, the conversion from binary to hexadecimal groups four digits of binary numbers to obtain a single hexadecimal digit. Four digits of a binary number (0000 to 1111) can represent a single digit in base 16 from 0 to F. A binary number is divided into one four-digit binary number or multiple four-digit binary numbers starting from the rightmost digit to the left-most digit. If the remaining number of digits is less than four in the division process, the 0(s) is added to the left of the remaining digit(s) to compose the four digits.

Example 1.15: Convert 1011110110_2 to base 16.

$10\ 1111\ 0110_2$

00 is added to the leftmost of 10, which are the leftmost two digits.

=> $0010\ 1111\ 0110_2$

2_{16} F_{16} 6_{16}

So, $1011110110_2 = 2F6_{16}$

Hexadecimal to Binary Conversion

Similarly, a digit of a hexadecimal number is represented by four digits of a binary number. If the number of digits of a binary number is less than four, 0(s) should be added to the leftmost position(s) of the binary number. Only 0(s) at the leftmost position(s) of the converted binary number can be waived.

Example 1.16: Convert $2B_{16}$ to base 2.

$2\ B_{16}$

=> $10_2\ 1011_2$

So, $2B_{16} = 101011_2$.

Conversion Between Octal and Hexadecimal

The base-8 (2^3) numbering system, octal, uses the digits 0 to 7 to represent numbers. The base-16 (2^4) numbering system, hexadecimal, uses the digits 0 to 9 and the letters A to F to represent numbers. Binary numbers are used as an intermediary to build the connection between octal and hexadecimal numbers.

Octal to Hexadecimal Conversion

Binary numbers are used as the medium in the conversion from octal to hexadecimal. Each unit digit of an octal number is first converted to a three-digit binary number. If the converted binary number has less than three digits, 0(s) should be added to the leftmost position of the converted binary number. Then the obtained three-digit binary number(s) are concatenated and divided into one or multiple four-digit binary numbers representing hexadecimal digits. Similarly, 0(s) can only be added to the leftmost position of the concatenated binary number. Finally, one four-digit binary number or each multiple four-digit binary number is converted to a hexadecimal digit.

Example 1.17: Convert 72_8 to base 16.

$7\ 2_8$

=> $111_2\ 010_2$ (grouping to convert octal to binary)

=> $11_2\ 1010_2$ (grouping to convert binary to hexadecimal)

=> $3_{16}\ A_{16}$ (convert binary groups to hexadecimal digits)

So, $72_8 = 111010_2 = 3A_{16}$.

Hexadecimal to Octal Conversion

Similarly, binary numbers are used as the medium in the conversion from hexadecimal to octal. Each unit digit of a hexadecimal number is first converted to a four-digit binary number. If the converted binary number has less than four digits, 0(s) should be added to the leftmost position of the converted binary number. Then the obtained four-digit binary number(s) are concatenated and divided into multiple three-digit binary numbers, each of which represents an octal digit. Similarly, 0(s) can only be added to the leftmost position of the concatenated binary number. Finally, the three-digit binary numbers are converted to octal digits.

Example 1.18: Convert $A2_{16}$ to base 8.

$A\,2_{16}$

=> $1010_2\ 0010_2$ (grouping to convert hexadecimal to binary)

=> $10_2\ 100_2\ 010_2$ (grouping to convert binary to octal)

=> $2_8\ 4_8\ 2_8$ (convert binary groups to hexadecimal digits)

So, $A2_{16} = 10100010_2 = 242_8$.

Arithmetic Operations of Binary Numbers

An n-bit binary number can represent any unsigned integer from 0 to $2^n - 1$. For example, 4 bits can represent the decimal values 0 to 15 ($2^4 - 1$) (as shown in Table 1.1); 8 bits can represent the values 0 to 255 ($2^8 - 1$). A given number of bits can only represent an allowable range of values. When a result value is out of the allowable range of values during arithmetic operations on binary numbers, overflow occurs. The largest number represented by 4 bits is 1111_2 (15_{10}). If any unsigned number greater than 0 is added to 1111_2, the sum will be greater than 1111_2 (causing overflow), because it cannot be represented by 4 bits.

Binary Number Addition

Binary number addition is similar to decimal number addition. The digits in the corresponding columns are added. A carry may be involved in adding the numbers in the same column. Table 1.3 shows all possible columns of addition results in binary.

A and B are the two values added; Carry-In is a third value, which represents the carry into the column; Sum is the column sum; and Carry-Out is a carry to the next column.

TABLE 1.3 Carry-In and Carry-Out Results in Binary Addition

A	B	CARRY-IN	SUM	CARRY-OUT
0	0	0	0	0
0	0	1	1	0
0	1	0	1	0
0	1	1	0	1
1	0	0	1	0
1	0	1	0	1
1	1	0	0	1
1	1	1	1	1

For example, add 10011010_2 to 1011101_2. Two unit digits involve a carry. The corresponding two carries are marked in red.

```
        1 1
  (1 0 0 1  1 0 1 0)2
+ (  1 0 1  1 1 0 1)2
  (1 1 1 1  0 1 1 1)2
```

Decimal number addition is used to demonstrate how carry-in works. Compare with binary number addition; carry-in plays the same role in decimal number addition. For example, $16_{10} + 25_{10} = 41_{10}$.

```
    1
    1 6
+   2 5
    4 1
```

Binary Number Subtraction

Binary number subtraction is similar to decimal number subtraction. The digits in the same column are subtracted. A borrow is needed when 1 is subtracted from 0. Table 1.4 shows all possible column subtractions given A − B.

TABLE 1.4 Borrows in Binary Subtraction

A	B	BORROW	DIFFERENCE
0	0	0	0
0	1	1	1
1	0	0	1
1	1	0	0

For example, 10110010_2 subtracts 1011101_2.

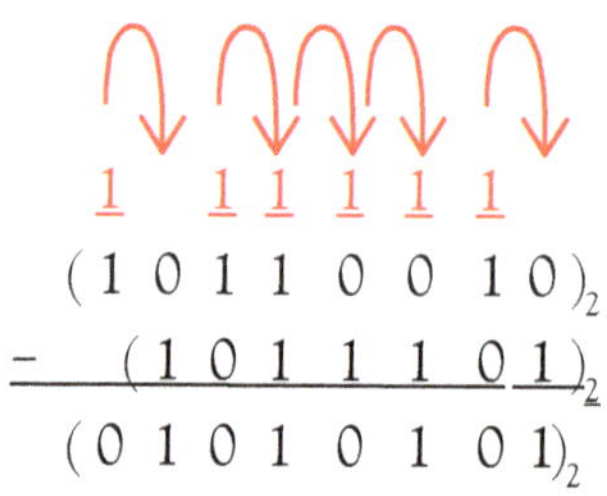

Decimal number subtraction is used to demonstrate how borrow works. Compare with binary number subtraction; borrow plays the same role in decimal number subtraction.

For example, $52_{10} - 37_{10} = 15_{10}$.

Chapter Summary

Computers are developed to process all types of data, including numerical data, text characters, and other multimedia information. Binary data is the foundation of the computer system. After reviewing the positional weighting in the well-known decimal numbering system, the binary numbering system is introduced.

The division-remainder method is introduced to convert a decimal number to a binary number. Each division by 2 would get a quotient and a remainder. The remainder is either 0 or 1. The process of dividing resulting quotients continues until the quotient reaches 0. After the successive divisions, the list of obtained remainders in a reverse order gives the conversion result. This division-remainder method is generic so that it can be used to convert a decimal number to any other base number, such as octal and hexadecimal.

The method of converting a fractional decimal number to a binary number is different. The fractional decimal number is repeatedly multiplied by 2. After each multiplication, an integer part of the result is taken as part of the final result. Note: The converted fractional binary data may not have the exact same value as its decimal counterpart. At this moment, an approximated value is used according to the bit size.

Computers only recognize binary data. Octal and hexadecimal numbers provide more human-friendly representations and are much easier to express binary number representations than in any other base number system. Conversion among binary, octal, and hexadecimal numbers is straightforward. Group four digits of binary numbers to obtain a single hexadecimal digit, and group three digits of binary numbers to obtain a single octal digit, and vice versa. Table 1.5 summarizes the basic conversions among binary, octal, decimal, and hexadecimal numbers.

Finally, basic addition and subtraction of binary numbers are introduced. Similar to regular operations for decimal numbers, carries and borrows could happen when adding and subtracting are performed.

TABLE 1.5 Conversion Among Binary, Octal, Decimal, and Hexadecimal Numbers

HEX	DECIMAL	OCTAL	BINARY
0	0	0	0
1	1	1	1
2	2	2	10
3	3	3	11
4	4	4	100
5	5	5	101
6	6	6	110
7	7	7	111
8	8	10	1000
9	9	11	1001
A	10	12	1010
B	11	13	1011
C	12	14	1100
D	13	15	1101
E	14	16	1110
F	15	17	1111
10	16	20	10000
20	32	40	100000
40	64	100	1000000
80	128	200	10000000
100	256	400	100000000
200	512	1000	1000000000
400	1024	2000	10000000000

Chapter 1 Exercises

1. Convert following numbers in different bases to base 10. Show your work.

    a. 31240_5
    b. 1552_6
    c. 112012_3
    d. 7213_8

2. Translate the following base 2 numbers to base 10. Show your work.

    a. 10110
    b. 1000000
    c. 1111
    d. 10101110
    e. 1011101
    f. 111111

3. Explain how you can quickly compute the base-10 value of 10000002 given your answer to 2f.

4. Translate the following base-10 numbers to base 2 using the Division-Remainder Algorithm. Show your work.

    a. 128
    b. 231
    c. 400
    d. 6593

5. Translate the following base-2 fractional numbers to base 10. Show your work.

    a. 0.1011
    b. 0.00001
    c. 0.111
    d. 0.101011

6. Translate the following base-10 fraction to a 5-bit base-2 fraction. Show your work.

   a. 0.3125
   b. 0.15625
   c. 0.53125
   d. 0.73

7. Compute the round-off error for 6d. Show your work.

8. Translate the following base-10 numbers to base 2. Use 4 bits after the decimal point. Show your work.

   a. 45.63
   b. 12.19

9. Compute the round-off error for 8b. Show your work.

10. Translate the following octal (base 8) numbers to binary.

    a. 37
    b. 473
    c. 2645
    d. 41132

11. Translate the following binary number to octal.

    a. 101011
    b. 10111011
    c. 1000110111011
    d. 1101010011011100101

12. Convert the following base-16 (hexadecimal) numbers to binary.

    a. 6FB
    b. 29AC5
    c. ABCDEF41

13. Convert the following binary numbers to hexadecimal.

    a. 1101010111111001101101
    b. 11101010001100110010101011011101011 1
    c. 101111010101111010110010101011110011011 10

14. Translate the base-10 number 73645 into binary, octal and hexadecimal.

15. Perform the following binary additions. Show your work.

    a. 100101 + 000110
    b. 101101 + 011011
    c. 11010010 + 10010110
    d. 10011001 + 10011110

16. Translate the two addends and sum from problem 15d to base 10. Show your work.

17. Convert the two operands and sum from problem 15d to base 10. Show your work.

18. Convert the two operands and difference from problem 17d to base 10. Show your work.

CHAPTER

2

Signed Integer Representation

Introduction

Integers include positive numbers, negative numbers, and zero. In computer systems, there are two categories of integer representation schemes (data types): unsigned and signed.

- Unsigned integers: can represent zero and positive integers.
- Signed integers: can represent zero, positive, and negative integers.

In Chapter 1, we have seen how computers convert unsigned integers into binary values without considering their positive or negative signs. In this chapter, we will discuss three different methods for representing positive and negative numbers: signed magnitude, one's complement, and two's complement.

Objectives

This chapter will help students:

- **understand** the relationship between signed and unsigned integer data types in programming languages.
- **understand** data size for different data types.
- **master** signed integer representation methods.

- **identify** the values of a binary number in two's complement system.
- **explain** the root cause of overflow.
- **apply** the detection approach and solution for monitoring overflow.

Key Terms

MSB: within a given data space, the bit in the leftmost position has the maximum weight. It is called most significant bit (MSB). For a signed data type, the sign bit is located in this position.

unsigned integer: a data type that typically represents the non-negative values, 0 to its max. There is no sign bit within the given space.

signed integer: a data type that typically represents negative, zero, and positive numbers. There is a sign bit within the given space, which is located at the leftmost position, called MSB.

overflow: in a computer system, a data variable has its fixed memory space according to its data type. If a variable receives data that exceeds its assigned data space, overflow happens, which would cause unexpected computational errors.

Unsigned and Signed Integers

Computers use a fixed number of bits to represent an integer. In C/C++ or other programming languages, there are several integer data types with different bit-lengths:

- char 8-bit
- unsigned char 8-bit
- short 16-bit
- unsigned short 16-bit
- int 32-bit
- unsigned int 32-bit
- long 64-bit
- unsigned long 64-bit

The data type with the keyword "unsigned" in the declaration is the **unsigned integer**. All associated bits are used for representing the number. For example, an unsigned short data variable has a 16-bit space, and its data range is 0 to 65,536:

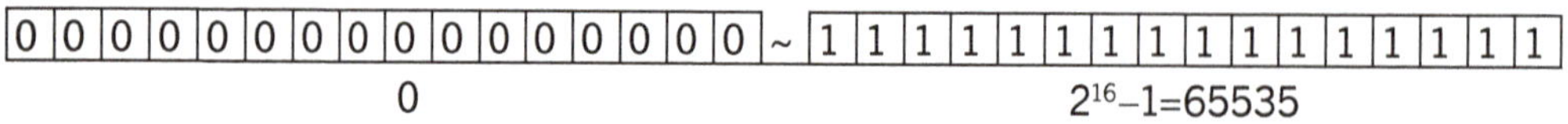

An unsigned integer can be used when we know that the value will always be non-negative. For example, if we are counting the number of players in a game, we could use an unsigned integer because there will always be 0 or more players.

The following C++ code shows how an unsigned short variable *s* is declared and used.

```
#include <iostream>
using namespace std;
int main()
{
    unsigned short s;
    s = 315;
    cout << “s=” << s<< endl;
    return 0;
}
```

Without the keyword "unsigned" in the declaration statement, the integer is the signed data type. **Signed integer** representation requires additional information: a special sign bit in the most significant (leftmost) position is used to indicate a positive or negative number. For example, if a 16-bit short data type integer is declared in a C/C++ program:

short *b*; //integer *b* with 16-bit space is allocated for this variable, and the leftmost bit is the sign bit

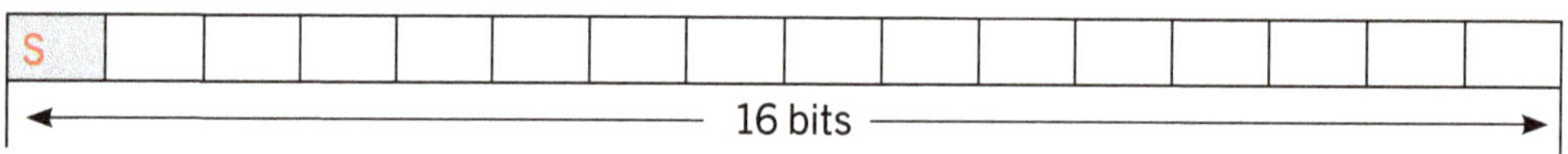

A "1" in the sign bit indicates a negative number. A "0" in the sign bit indicates a positive number. The remaining 15 bits represent the absolute value of the number.

Since the absolute value of a number is determined by the remaining bits of the given space, if the total size is *n* bits, the range of numbers that can be represented is from $-(2^{n-1} - 1)$ to $+(2^{n-1} - 1)$.

For example, given a 16-bit space (e.g., *short* in C/C++ language), the maximum value it can represent is positive (sign bit=0) with the largest absolute value determined by the remaining 15 bits. That is:

$$2^{15} - 1 = 32{,}767$$

Similarly, the smallest value is negative (sign bit=1) with the largest absolute value:

$$-(2^{15} - 1) = -32{,}767$$

Hence, the data range of a 16-bit integer is [–32,767, 32,767].

Table 2.1 shows the sizes and ranges of some integer data types.

TABLE 2.1 Bit Size and Data Range of Integer Data Types in C/C++ Language

DATA TYPE (C/C++)	DATA SIZE (BITS)	RANGE IN DECIMAL
char	8	–128 to +127
unsigned char	8	0 to 255
short	16	–32,767 to +32,767
unsigned short	16	0 to 65,536
int	32	–2,147,483,647 to +2,147,483,647
unsigned int	32	0 to 4,294,967,295
long	64	9,223,372,036,854,775,807 to 9,223,372,036,854,775,807
unsigned long	64	0 to 18,446,744,073,709,551,615

With the exception of the leftmost sign bit, how the remaining part of a signed integer number is organized depends on the method used. There have been three methods in history: signed magnitude, one's complement, and two's complement, which are introduced in the following sections.

Signed Magnitude (SM)

Signed magnitude (SM) is the most intuitive and simplest method of representing signed integers. In this method, the most significant bit (MSB) is a sign bit, and the remaining digits are the magnitude bits representing the absolute value.

- If the MSB is a 0, it indicates it is a positive number.
- If the MSB is a 1, this indicates that the number is negative.
- The other bits represent the magnitude (absolute value) of the number.

Example 2.1

Given 8-bit space, consider the SM notation for representing 3 and –3:

- The sign bit (MSB) of 3 is 0 and the sign bit of –3 is 1.
- The absolute values of both are the same $(3)_{10} = (11)_2$.
- The values in the remaining 7 bits should be: 0000011 (absolute value padded left with zeros).
- Thus, the bit patterns are:

3:	0	0	0	0	0	0	1	1
–3:	1	0	0	0	0	0	1	1

Signed magnitude notation makes arithmetic operations a bit complicated. We use the following example to illustrate the complexity.

Example 2.2

Given an 8-bit data space, using the signed magnitude method, compute the following two cases:

a. – 46 – 25
b. 46 – 25

- Get the 7-bit binaries of the absolute values from both cases:

$$46 = (010\ 1110)_2 \quad 25 = (001\ 1001)_2$$

- With the sign bit, the SM representations for –46, –25, and 46 are:

(–46) = (1010 1110), –25 = (1001 1001), 46 = (0010 1110)

- For case (a), (–46) + (–25), since the signs of both numbers are the same, the system does addition on their magnitude bits:

1		010 1110	(–46)
1	+	001 1001	(–25)
1		100 0111	(–71)

For the sign bit, just take the existing sign 1 into the final without performing an adding operation. Thus, the result in signed magnitude is $(1100\ 0111)_2$, which is –71 in decimal.

- For case (b): 46 + (–25), since the signs of the numbers are different, the system just does regular **subtraction** on the magnitude bits and uses the larger absolute value as the minuend, and the smaller one as the subtrahend. Specifically, for this case, it is 46 – 25:

0		010 1110	(46)
1	–	001 1001	(–25)
0		001 0101	(21)

For the sign bit, just use the sign from the number with the larger absolute value. Here, 46 is larger; thus, the positive sign 0 is taken for the result. Thus, the result in signed magnitude is $(0001\ 0101)_2$, which is 21 in decimal.

From **Example 2.2**, we see that, by considering the conditions of the sign bits from both operands and their absolute values, the system determines whether addition or subtraction is performed and whether the sign bit should be 0 or 1. This extra processing complicates arithmetic operations with SM numbers.

Signed magnitude suffers another disadvantage of having two different representations for zero: positive zero and negative zero. The following are the zeros in an 8-bit space:

Positive zero:	0	0	0	0	0	0	0	0
Negative zero:	1	0	0	0	0	0	0	0

Table 2.2 shows some signed decimal numbers and their equivalent in SM notation given the size of 8 bits.

TABLE 2.2 Some Integer Examples and Their Signed Magnitude Representations

SIGNED DECIMAL	SIGNED MAGNITUDE
+6	0000 0110
−6	1000 0110
+0	0000 0000
−0	1000 0000
+7	0000 0111
−7	1000 0111

Although signed magnitude representation is easy for people to understand, it is seldom used in practice due to these disadvantages. Computer systems use complement notations for signed value representation.

One's Complement

One's complement is another way to represent signed integers in computer systems. In this method:

- The most significant bit (MSB) is the sign bit, "0" for representing positive and "1" for negative integers.
- The remaining $n - 1$ bits represent the magnitude (absolute) of the integer, as follows:
  - For a positive integer, the absolute value is equal to the magnitude of the remaining $n - 1$ bits.
  - For a negative integer, the absolute value is equal to the magnitude of the complement (inverse, all 0s changed to 1s and 1s changed to 0s) of the $(n - 1)$-bit binary pattern (hence called one's complement).

Figure 2.1 shows the flowchart of converting a decimal integer into binary in one's complement. For a positive number, its one's complement representation is the same as signed magnitude. For a negative number, it converts its absolute value into binary and inverts every bit.

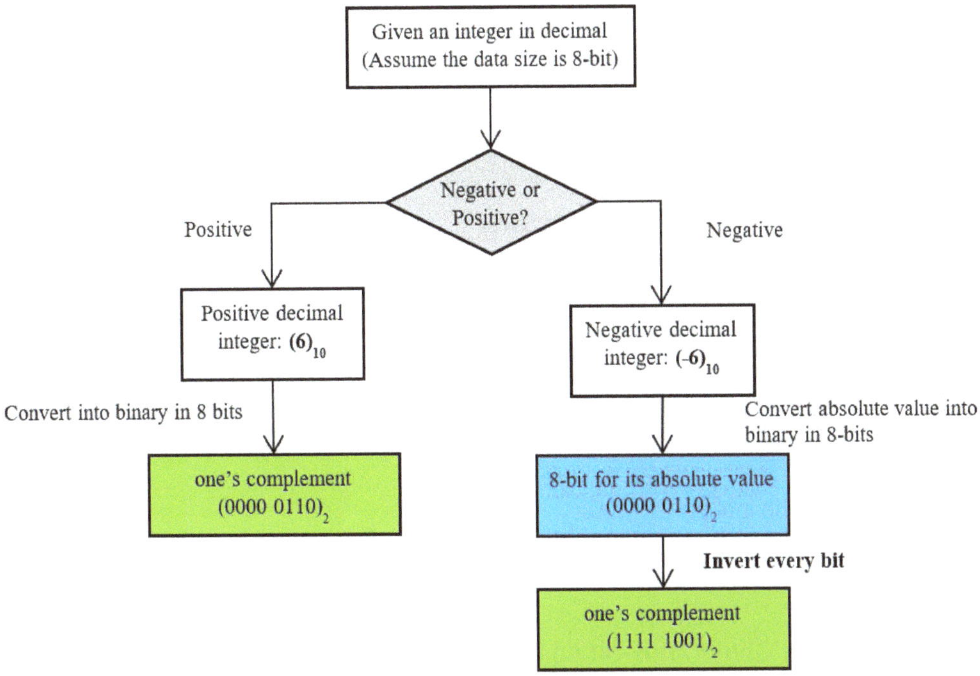

FIGURE 2.1 Flowchart Representing a Decimal Integer in One's Complement

Example 2.3: Given an 8-bit space, show the bit pattern of $(6)_{10}$ using one's complement notation.

- Since it is a positive number, its sign bit is 0.
- Convert its absolute value into binary $(6)_{10} = (0000\ 0110)_2$.
- The leftmost bit is 0:

0	0	0	0	0	1	1	0

Example 2.4: Given an 8-bit (n = 8) one's complement representation, show the decimal value.

0	1	0	0	0	0	0	1

- Sign bit is 0, so it is a positive number.
- Its absolute value is determined by the remaining 7 bits: $(100\ 0001)_2 = (65)_{10}$.
- Hence, the integer is $+(65)_{10}$.

Note: The representation of positive numbers in one's complement is the same as in signed magnitude notation.

Example 2.5: Given an 8-bit ($n = 8$) one's complement representation, show the decimal value.

1	0	0	0	0	0	0	1

- Sign bit is 1, so it is a negative number.
- Its absolute value is the complement of every bit (invert every bit):

1	0	0	0	0	0	0	1
↓	↓	↓	↓	↓	↓	↓	↓
0	1	1	1	1	1	1	0

$(0111\ 1110)_2 = (126)_{10}$

- Hence, the integer is $(-126)_{10}$.

Example 2.6: Given an 8-bit space, show the bit pattern of $(-3)_{10}$ in one's complement notation.

- Since it is a negative number, its sign bit is 1
- Convert the absolute value into binary in 8-bit space $(3)_{10}=(0000\ 0011)_2$
- Invert this 8-bit pattern (flip over every bit):

0	0	0	0	0	0	1	1
↓	↓	↓	↓	↓	↓	↓	↓
1	1	1	1	1	1	0	0

- For negative one's complement, remember the sign bit is 1.

With one's complement, the complexity of arithmetic operations is reduced compared with the signed magnitude method because the appropriate sign bit will be generated without a need for special conditional tests. Additionally, the hardware digital circuits for the subtraction operation can be eliminated, as any subtraction operation can be replaced with adding a negative number. After adding, the carry-out is added to the resulting sum to compute the final one's complement value. Note that the carry-out may be 0, meaning that there will be no change to the computed sum. We use the following example to illustrate subtraction in one's complement notation.

Example 2.7: Using one's complement binary arithmetic (8-bit), compute 48 – 19.

- $(48)_{10}$ is positive; its one's complement representation is:

0	0	1	1	0	0	0	0

- $(-19)_{10}$ is negative; convert its absolute value into binary in 8-bit space:

$$(19)_{10} = (0001\ 0011)_2$$

- Invert every bit:

0	0	0	1	0	0	1	1
↓	↓	↓	↓	↓	↓	↓	↓
1	1	1	0	1	1	0	0

- Thus, –19 in one's complement within 8-bit space is:

1	1	1	0	1	1	0	0

- Internally, 48 + (–19) is performed:

$$\begin{array}{r} 0011\ 0000 \\ +\ \underline{1110\ 1100} \\ \boxed{1}\,0001\ 1100 \end{array}$$

- Please note: There is a carry-out $\boxed{1}$ exceeding the given 8-bit space. This carry-out should be added to the result:

$$\begin{array}{rl} 0001\ 1100 & \\ +\ \underline{\qquad\qquad \boxed{1}} & \text{(carry-out)} \\ 0001\ 1101 & \end{array}$$

- Thus, the result is $(0001\ 1101)_2 = (29)_{10}$.

Compared with signed magnitude, the arithmetic operation in one's complement is more straightforward, as no subtraction or conditional operations are needed. However, there is an extra step of adding the carry-out exceeding the space to the result.

Another disadvantage of one's complement is that we still have two representations for 0. The following two examples show the representations for positive and negative zeros.

Example 2.8

An 8-bit binary representation with all zeros:

0	0	0	0	0	0	0	0

- It is a positive number as the sign bit is 0.
- Its absolute value is $(000\ 0000)_2 = (0)_{10}$.
- Hence, the integer is $(+0)_{10}$.

Example 2.9

An 8-bit binary representation with all ones:

1	1	1	1	1	1	1	1

- It is a negative number as the sign bit is 1.
- Its absolute value is the complement of the remaining 7 bits:

1	1	1	1	1	1	1	1
↓	↓	↓	↓	↓	↓	↓	↓
0	0	0	0	0	0	0	0

$(0000\ 0000)_2 = (0)_{10}$

- Hence, the integer is $(-0)_{10}$.

Because of this limitation of the one's complement model shown in the examples, computer engineers long ago stopped using one's complement in favor of the more efficient two's complement representation for signed integers.

Two's Complement

Modern computers use the two's complement method to represent signed integers. In this representation:

- Again, the most significant bit (MSB) is the sign bit, with value of 0 representing positive integers and 1 representing negative integers.
- The remaining $n - 1$ bits represent the magnitude of the integer, as follows:
  - For a positive integer, the representation is the same as the signed magnitude and one's complement. That is, the magnitude of the $(n - 1)$-bit binary pattern presents the positive value of the integer.

- For a negative integer, the absolute value of the integer is equal to "the magnitude of the complement of the (n – 1)-bit binary pattern plus one" (hence why it's called two's complement).

Figure 2.2 shows the flowchart of converting a decimal integer into binary in two's complement.

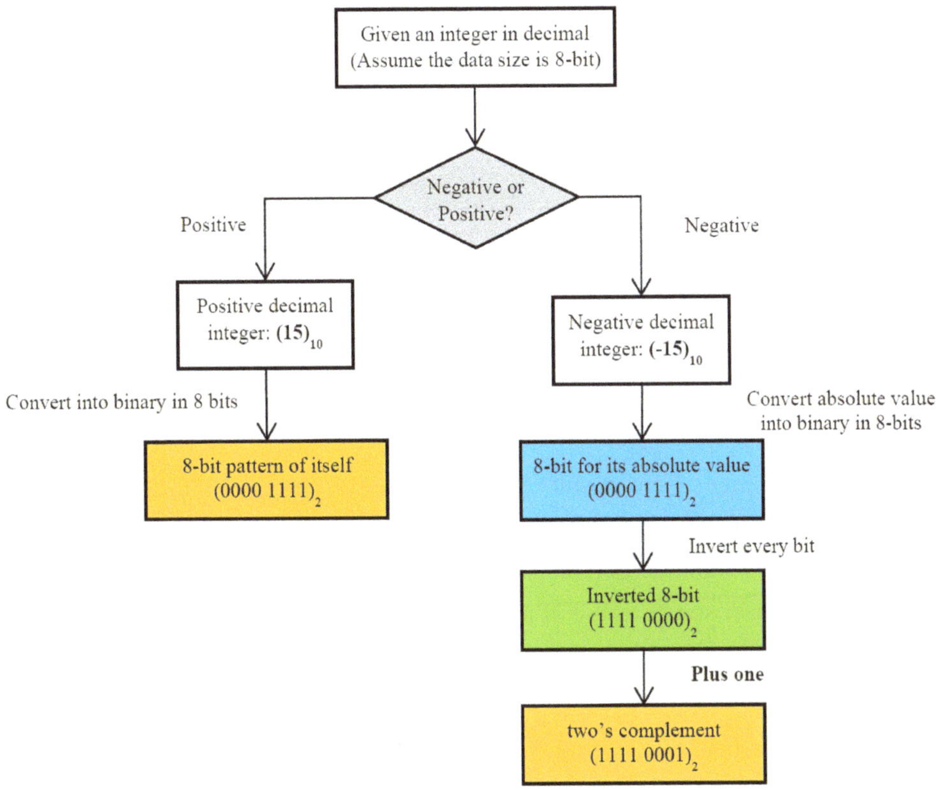

FIGURE 2.2 The Flowchart of Conversion for a Decimal Integer in Two's Complement Representation

Example 2.10: Given an 8-bit space, show the bit pattern of $(16)_{10}$ using two's complement notation.

- Since it is a positive number, directly convert to binary within the given space: $(16)_{10} = (0001\ 0000)_2$

0	0	0	1	0	0	0	0

- Note: The sign bit is 0 for a positive number.

Example 2.11: Given an 8-bit space, show the bit pattern of $(-13)_{10}$ using two's complement notation.

- Since it is a negative number, convert the absolute value into binary:

$$(13)_{10} = (0000\ 1101)_2$$

- Invert every bit of this 8-bit pattern:

0	0	0	0	1	1	0	1
↓	↓	↓	↓	↓	↓	↓	↓
1	1	1	1	0	0	1	0

- Plus 1:

1	1	1	1	0	0	1	0	+	1	=	1	1	1	1	0	0	1	1

- Note: The sign bit is 1 for a negative number.

Given a sequence of binary bits in two's complement representation, Figure 2.3 shows the flowchart of converting it to decimal. In the negative case, the order of the operations is similar to converting decimals to binaries: by inverting every bit and adding 1, the absolute value of a negative number can be obtained.

Example 2.12

Given an 8-bit space, using two's complement, what is the decimal value of the bit pattern:

1	1	1	0	1	0	1	1

- It is negative as the sign bit is 1. Then, go to the right branch of the flowchart.
- Invert every bit

1	1	1	0	1	0	1	1
↓	↓	↓	↓	↓	↓	↓	↓
0	0	0	1	0	1	0	0

- Plus 1:

0	0	0	1	0	1	0	0	+	1	=	0	0	0	1	0	1	0	1

- The absolute value of the result:

$$(0001\ 0101)_2 = (21)_{10}$$

- Given the initial sign bit of 1, the decimal value is $(-21)_{10}$.

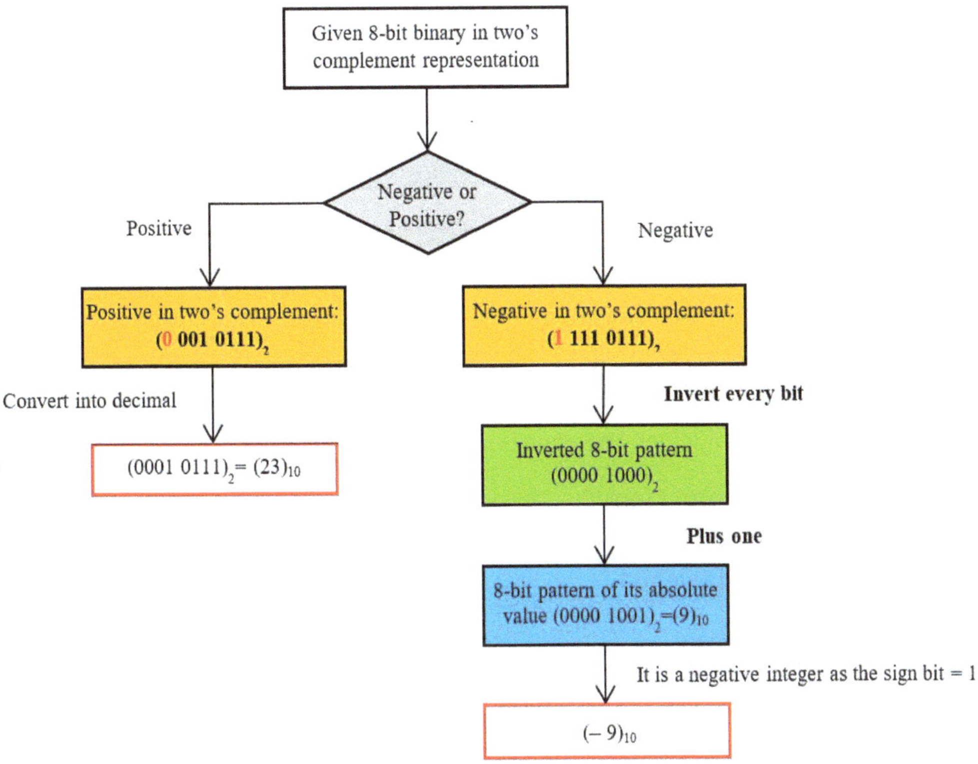

FIGURE 2.3 Flowchart of Conversion for a Given Two's Complement Representation into Decimal

With two's complement, arithmetic operations are simpler than one's complement. When performing subtraction, it just adds a negative integer and discards the carry-out exceeding the space. We use the following examples to illustrate subtraction in two's complement notation. Note when adding a two's complement number that the leading bit is treated as part of a number and not as a sign bit like signed magnitude notation. It is involved in the addition/subtraction like any other bit.

Example 2.13: Using two's complement notation in 8-bit space, compute 28 – 25.

- $(28)_{10}$ is positive; its two's complement representation is:

0	0	0	1	1	1	0	0

- $(-25)_{10}$ is negative; convert its absolute value into binary:

$$(25)_{10} = (0001\ 1001)_2$$

- Invert every bit of the absolute value:

0	0	0	1	1	0	0	1
↓	↓	↓	↓	↓	↓	↓	↓
1	1	1	0	0	1	1	0

- After adding 1, the two's complement of –25 is:

1 1 1 0 0 1 1 0 + 1 = 1 1 1 0 0 1 1 1

- Internally, the addition operation is performed 28 + (–25):

```
   0 001 1100
+  1 110 0111
 ------------
 1 0 000 0011
```

- Please note: There is a carry-out 1 exceeding the given 8-bit space, which is discarded, and the result is: $(0000\ 0011)_2 = (3)_{10}$.

Example 2.14: Using two's complement notation in 8-bit space, compute 48 – 19.

- $(48)_{10}$ is positive; its two's complement representation is:

0	0	1	1	0	0	0	0

- $(-19)_{10}$ is negative; convert its absolute value into binary:

$$(19)_{10} = (0001\ 0011)_2$$

- Invert every bit in this 8-bit space:

0	0	0	1	0	0	1	1
↓	↓	↓	↓	↓	↓	↓	↓
1	1	1	0	1	1	0	0

- Plus 1, the bit pattern of –19 in two's complement is:

1	1	1	0	1	1	0	0	+	1	=	1	1	1	0	1	1	0	1	

- Internally, 48 – 19 is performed as 48 + (–19):

```
   0011 0000
+  1110 1101
-----------
  10001 1101
```

- Please note: There is a carry-out 1 exceeding the given 8-bit space. This carry-out is discarded, and the result is: $(0001\ 1101)_2 = (29)_{10}$.

Example 2.15: Using two's complement notation in an 8-bit space, compute –12 – 4.

- Both operands are negatives; the 8-bit binary patterns of their absolute values are:

12:	0	0	0	0	1	1	0	0
4:	0	0	0	0	0	1	0	0

- Invert the binary patterns of both absolute values:

12:	0	0	0	0	1	1	0	0
	↓	↓	↓	↓	↓	↓	↓	↓
	1	1	1	1	0	0	1	1

4:	0	0	0	0	0	1	0	0
	↓	↓	↓	↓	↓	↓	↓	↓
	1	1	1	1	1	0	1	1

- Plus 1:

1	1	1	1	0	0	1	1	+	1	=	1	1	1	1	0	1	0	0
1	1	1	1	1	0	1	1	+	1	=	1	1	1	1	1	1	0	0

- The bit patterns for both in two's complement are:

–12:	1	1	1	1	0	1	0	0
–4:	1	1	1	1	1	1	0	0

- Sum them:

```
     1111 0100    (–12)
+    1111 1100    (–4)
    11111 0000
```

(The leading carry-out 1 of the result is struck through.)

- Note: There is a carry-out of the sign bit, which is truncated. The result with sign bit=1 is in two's complement.
- Let us convert this negative result into decimal by following the flowchart in Figure 2.3:

  – Invert every bit:

1	1	1	1	0	0	0	0
↓	↓	↓	↓	↓	↓	↓	↓
0	0	0	0	1	1	1	1

  – Plus 1:

0	0	0	0	1	1	1	1	+	1	=	0	0	0	1	0	0	0	0

  – The added result is the absolute value:

$$(0001\ 0000)_2 = (16)_{10}$$

  – Given the initial sign bit of 1, the decimal value is $(-16)_{10}$

Overflow Issue in Arithmetic Operation

No matter what kind of representation the system uses, **overflow** happens if the given space cannot accommodate a value. We illustrate the overflow issue using the two's complement representation.

Example 2.16: In two's complement, given an 8-bit space, compute 107 + 46.

- Both 107 and 46 are positive; their two's complement representations are:

107:	0	1	1	0	1	0	1	1
46:	0	0	1	0	1	1	1	0

- Sum them:

	0	1	1	0	1	0	1	1	(107)
+	0	0	1	0	1	1	1	0	(46)
	1	0	0	1	1	0	0	1	

- There is a carry that flows into the sign bit, and the result is: $(1\ 001\ 1001)_2$.
- Note: The number is in two's complement, and the sign bit is 1. So, the system believes it is a negative number.
  - Invert every bit of the above result:

1	0	0	1	1	0	0	1
↓	↓	↓	↓	↓	↓	↓	↓
0	1	1	0	0	1	1	0

  - Plus 1:

0	1	1	0	0	1	1	0	+ 1 =	0	1	1	0	0	1	1	1		

  - The added result is the absolute value:

(	0	1	1	0	0	1	1	1	$)_2$	$= (103)_{10}$

- Since it is negative, the system believes the integer is $(-103)_{10}$

Hence, we have the erroneous result: 107 + 46 = –103. This phenomenon is called overflow.

The root cause is that the data space is not large enough to hold the value. The correct answer of this calculation should be 253. However, for the given 8-bit signed data, the maximum value that can be represented is +127. 253 is out of the data range and cannot be represented in 8-bit two's complement.

We see the result in **Example 2.16** is obviously wrong, adding two positive numbers and getting a negative result. However, the computer detects overflow by monitoring the carry-in and carry-out on the sign bit. In the previous example, we can notice that:

- There is a carry-in into the sign bit (make the bit 1 after summation).
- There is NO carry-out of the sign bit.

If the carry-in and carry-out of the sign bit is not equal, overflow has occurred. Let us see other overflow examples.

Example 2.17: In two's complement, given an 8-bit space, compute (–124 – 116).

- Both operands are negative; the 8-bit binary patterns of their absolute values are:

124:	0	1	1	1	1	1	0	0
116:	0	1	1	1	0	1	0	0

- Invert 8-bit binary patterns of both absolute values.

0	1	1	1	1	1	0	0
↓	↓	↓	↓	↓	↓	↓	↓
1	0	0	0	0	0	1	1

0	1	1	1	0	1	0	0
↓	↓	↓	↓	↓	↓	↓	↓
1	0	0	0	1	0	1	1

- Add 1 to each:

1	0	0	0	0	0	1	1	+	1	=	1	0	0	0	0	1	0	0
1	0	0	0	1	0	1	1	+	1	=	1	0	0	0	1	1	0	0

- The bit patterns of both numbers in two's complement are:

–124:	1	0	0	0	0	1	0	0
–116:	1	0	0	0	1	1	0	0

- Sum both numbers:

```
    1000 0100   (–124)
+   1000 1100   (–116)
   10001 0000
```

- Note: There is a carry-out of the sign bit, which is truncated. The sign bit within the 8-bit space is 0. Thus, the result is positive 16:

$$(0001\ 0000)_2 = (16)_{10}$$

- This is an erroneous result: $-124 - 116 \neq 16$

Again, the root cause of this overflow case is that the 8-bit space is not large enough for the correct result $(-240)_{10}$, since the smallest value that 8-bit two's complement can represent is -127.

We can also notice that:

- There is no carry-in flowing into the sign bit.
- But there is a carry-out of the sign bit.

The carry-in and carry-out of the sign bit are not equal. Thus, overflow has occurred. By watching the carry-in and carry-out of the sign bit, the system can detect overflow.

The following is an example showing balanced carry-in and carry-out on the sign bit.

Example 2.18: In two's complement method, given an 8-bit space, compute (23 – 9).

- The binary patterns of both absolute values are:

23:	0	0	0	1	0	1	1	1
9:	0	0	0	0	1	0	0	1

- For 23, its two's complement bit pattern is:

23:	0	0	0	1	0	1	1	1

- For –9, further invert every bit of its absolute value:

0	0	0	0	1	0	0	1
	↓	↓	↓	↓	↓	↓	↓
1	1	1	1	0	1	1	0

- Plus 1, –9 in two's complement is:

1	1	1	1	0	1	1	0	+	1	=	1	1	1	1	0	1	1	1

- Sum them 23 + (– 9):

```
      0001 0111    (23)
+     1111 0111    (–9)
     10000 1110
```

- From the 8-bit space, the sign bit value is 0. Thus, the result is a positive result:

$$(0000\ 1110)_2 = (14)_{10} \quad \rightarrow \quad 23 - 9 = 14$$

- Note: There is a carry-in into the sign bit and a carry-out of the sign bit (exceeding the data space).

In **Example 2.18**, we notice that:

- There is a carry-in into the sign bit.
- There also is a carry-out of the sign bit (exceeding the data space).
- The carries in and out of the sign bit are equal.

Thus, overflow has not occurred. By comparing the carry-in and carry-out of the sign bit, the overflow is detectable.

In summary:

- Overflow happens if the data space is not big enough for a value.
- The only solution to the overflow issue is increasing the data size. For example, using 16-bit space (short in C/C++) or other bigger data size can avoid above overflow problems.
- Overflow is detectable during the calculation by monitoring the carries on the sign bit. Specifically, when the carry-in into the sign bit is not equal to the carry-out, overflow has occurred. Otherwise, there is no overflow.
- Note: Overflow is not determined by only looking at the carry-out of the sign bit. A carry-out of 1 does not necessarily imply overflow, and a carry-out of 0 does not necessarily imply no overflow. The carry-in and carry-out must be compared.

Examples of Overflow Issue in C++ Programming

Given two positive decimal integers, 32,700 and 32,750, using a signed short data type in C++. Since two's complement is used, the bit patterns are:

$(32700)_{10}$	0	1	1	1	1	1	1	1	1	0	1	1	1	1	0	0
$(32750)_{10}$	0	1	1	1	1	1	1	1	1	1	1	0	1	1	1	0

If we add these numbers, the result should be 65,450. However, by executing the following C++ program, it presents a very strange number: -86.

```
#include <iostream>
using namespace std;
int main()
{
     short  a, b;
    a = 32700;
    b= a + 32750;
    cout << "b=" <<b<< endl;
    return 0;
}
```

Execution result:

```
$g++ -o main *.cpp
$main
b = -86
```

Why is it -86? Look at adding in detail. There is a carry-in into the sign bit and no carry-out (not equal):

```
  0111 1111 1011 1100    (32,500)
+ 0111 1111 1110 1110    (32,570)
  1111 1111 1010 1010
```

Remember, the result is in two's complement representation. Following the flowchart of Figure 2.3:

- The sum is negative given that the sign bit is 1, which leads to the right branch of the flowchart.
- Take the complement of the sum:

1	1	1	1	1	1	1	1	1	0	1	0	1	0	1	0
↓	↓	↓	↓	↓	↓	↓	↓	↓	↓	↓	↓	↓	↓	↓	↓
0	0	0	0	0	0	0	0	0	1	0	1	0	1	0	1

- Add 1:

0	0	0	0	0	0	0	0	0	1	0	1	0	1	0	1	+ 1 =	0	0	0	0	0	0	0	0	0	1	0	1	0	1	1	0	

- The absolute decimal value is:

0	0	0	0	0	0	0	0	0	1	0	1	0	1	1	0	=	$(86)_{10}$

- Considering the initial sign bit, the sum is $(-86)_{10}$:

As we know, the solution to the overflow is increasing the data size. Here, we change the data type to int (32 bits), and the result will be correct: $(65{,}450)_{10}$

```
#include <iostream>
using namespace std;
int main()
{
    int  a, b;
    a = 32700;
    b= a + 32750;
    cout << "b=" <<b<< endl;
    return 0;
}
```

Execution result:
$g++ -o main
*.cpp
$main
b = 65450

Chapter Summary

Integers include positive numbers, negative numbers, and zero. In the computer systems, there are two categories of integer representation (data types): unsigned and signed. For a signed integer data, there is a sign bit at the leftmost position indicating its sign. If the sign bit is 0, the number is positive. Otherwise, it is negative. An unsigned integer number is always non-negative, and it does not need a sign bit. Because of this, the data range of signed and unsigned variables is different.

For signed integers, there have been three representation methods in history: signed magnitude, one's complement, and two's complement. Signed magnitude (SM) notation is the most intuitive and simplest method of representing signed integers. But it makes arithmetic operations complicated since it requires the system to check conditions when performing calculations. This extra conditional test complicates arithmetic operations. Signed magnitude suffers another disadvantage of having two different representations for zero: positive zero and negative zero.

With one's complement, the complexity of arithmetic operations is reduced because the appropriate sign bit will be generated without a need for special conditional tests. Additionally, the hardware digital circuits for the subtraction operation can be eliminated since any subtraction operation can be replaced with adding a negative number. However, there is an extra step of adding the carry-out exceeding the space to the result. Another disadvantage of one's complement is that we still have two representations for zero.

Modern computers use the two's complement method to represent signed integers. With two's complement, arithmetic operations are simpler than with one's complement. When performing subtraction, it just adds a negative integer and discards the carry-out, exceeding the space without any conditional test.

Overflow happens if the data space is not big enough for a value. The only solution is increasing the data size. Overflow is detectable during the calculation by monitoring the carries on the sign bit. Specifically, when the carry-in into the sign bit is not equal to the carry-out, overflow has occurred. Otherwise, there is no overflow.

Chapter 2 Exercises

1. What decimal value does the 8-bit binary number 10011110 have if it is:

   a. an unsigned number?
   b. a signed magnitude number?
   c. a one's complement number?
   d. a two's complement number?

2. Translate the following numbers into 8-bit signed magnitude, 8-bit one's complement, and 8-bit two's complement notations respectively:

	SIGNED MAGNITUDE	ONE'S COMPLEMENT	TWO'S COMPLEMENT
4			
–4			
0			
7			
–18			
49			
19			
–19			

3. Perform the following calculation in 8-bit two's complement. Show the steps of your work (carries and sums). Additionally, indicate for each: (1) whether there was a carry-out; (2) whether there was overflow.

   a. 28 - 25
   b. 58 - 20
   c. 105 + 45
   d. -12 - 4
   e. 108 + 32
   f. 15 - 16

4. How do we detect overflow? Give an example.

5. How do we solve the overflow problem? Explain your answer.

CHAPTER

3

Floating Point Representation

Introduction

Integers and floats are two different kinds of numerical data. An integer is a number without a decimal point. A float is a floating-point number, which means it is a number that has a decimal place. Floats are used when more precision is needed. Chapter 2 explains the integer data representation in computer systems. How do computers recognize and represent floating-point numbers? Previously, different hardware manufacturers developed their own ways to implement floating point representations. It caused data incompatibility problems during data transportation and communication due to diverse implementations. To overcome this issue, IEEE provides a technical standard, IEEE 754, to regulate the floating number formats that many hardware manufacturers follow.

Objectives

This chapter will help students:

- **understand** scientific notation.
- **explain** the necessity of the IEEE 754 standard.
- **master** the skill of converting between decimal and floating-point binary representations (single precision).
- **understand** the difference between single precision and double precision in the IEEE 754 standard.

Key Terms

mantissa: A real number contains an integer part and a fractional part. The fractional part is called the mantissa.

significand: part of a number in scientific notation or in floating-point representation. The significand may represent an integer or a fraction.

binary normalized form: a form required by IEEE 754 to represent a binary floating-point number. It has a leading 1 and scaled by an exponent in base 2: $(1.\text{mantissa})_2 \times 2^{exp}$.

Scientific Notation

A real number can contain an integer part and a fractional part (called the mantissa). For example,

$$-315.28$$

−315 is the integer, and 0.28 is the mantissa.

A real number can also be represented in a scientific notation with a number of significant digits (the significand) and scaled using an exponent in some fixed base. The base for the scaling is normally 10 for decimals and 2 for binaries. Thus, a real number can be represented exactly as in the following form:

$$\text{significand} \times \text{base}^{\text{exponent}}$$

where significand has both integer and mantissa parts, **base** is an integer greater than or equal to 2, and the **exponent** is also an integer. The following is the more detailed form:

$$\underbrace{\text{integer. mantissa}}_{\text{significand}} \times \text{base}^{\text{exponent}}$$

For example, $\pi = 3.1415926$, can be represented as:

$$\underbrace{314.15926}_{\text{significand}} \times 10^{-2}$$

- Integer: 314
- Mantissa: 0.15926
- Base: 10
- Exponent: −2

This representation of a real number is not unique because it can also be represented as:

$$\underbrace{0.31415926}_{\text{significand}} \times 10^{1}$$

- Integer: 0
- Mantissa: 0.31415926
- Base: 10
- Exponent: 1

For a single real number, in the **significand** part, there is no fixed number of digits before and after the radix point (decimal point); that is, the radix point can "float."

The binary numbers in the computer system share a similar property.

For example, $(101.011)_2$ can be represented as:

$$(101.011)_2 = (10.1011)_2 \times 2^1$$

- Integer: $(10)_2$
- Mantissa: $(0.1011)_2$
- Base: 2
- Exponent: 1

Or

$$(101.011)_2 = (1.01011)_2 \times 2^2$$

- Integer: $(1)_2$
- Mantissa: $(0.01011)_2$
- Base: 2
- Exponent: 2

We can see that, with different exponent values, the decimal point is moving. Thus, the term "floating-point" is derived and can be thought of as a kind of scientific notation. The exponent indicates how far the radix point "floats" to derive the number in standard format. If the exponent is positive, the radix point floats to the right. If the exponent is negative, the radix point floats to the left. This is true for all base values, including base 2. By defining the form of the significand (integer and mantissa parts) of a floating-point number, computers are capable of representing real numbers.

IEEE 754 Floating-Point Standard

In 1985, the Institute of Electrical and Electronic Engineers (IEEE) published a floating-point standard for both 32-bit single- and 64-bit double-precision floating-point numbers. This standard is officially known as IEEE 754 (1985). Modern computers adopt the IEEE 754 standard for representing floating-point numbers. It was revised in 2008 as IEEE 754 (2008). This standard is adopted by almost all modern machines. There are two types of floating-point numbers defined in this standard: single-precision floating-point and double-precision floating-point numbers.

32-bit Single-Precision Floating-Point Numbers

The IEEE 754 32-bit single-precision floating-point numbers consist of three parts:

- 1 sign bit, 0: positive, 1: negative;
- 8 bits for the exponent part, using excess-127 bias (representing the exponent on a power of 2);
- 23 bits for the fractional part called a mantissa.

Sign	Exp bias	Mantissa
1 bit	⟵ 8 bits ⟶	⟵-----------23 bits -----------⟶

FIGURE 3.1 The IEEE 754 32-bit Single-Precision Floating-Point Representation

The structure of this 32-bit model is presented in Figure 3.1. The sign bit is at the most significant position. The floating-point number is positive or 0 when the sign bit is 0. Otherwise, the sign bit is 1.

In the IEEE 754 model, the significand is normalized with an implicit leading 1 and scaled by an exponent in base 2:

$$(1.\text{mantissa})_2 \times 2^{\text{exp}}$$

This is the so-called "**binary normalized form**" or "normalized form." For a decimal real number 7.0, first convert it into binary, and then normalize it as:

$$(7.0)_{10} = (111.0)_2$$

$$= (1.11)_2 \times 2^{2}$$

Here, the mantissa is 0.11 in binary, and the actual exponent is 2 in decimal. Note that the leftmost 1 is assumed and thus is not included in the mantissa.

Under the normalized form, the exponent value ranges from 1 to 254 in decimal where their corresponding 8-bit patterns are 00000001 and 11111110, respectively. The exponent part uses excess-127 bias. Given the actual exponent 2, after applying bias-127, the *Exp bias* is 129.

$$\begin{cases} exp = 2 \\ \text{exp} = \text{Exp bias} - 127 \end{cases} \quad \Rightarrow \quad \text{Exp bias} = 129$$

Thus, the 8-bit exponent part should be the bit pattern of 129: $(129)_{10} = (10000001)_2$.

Finally, put 0 into the sign bit, $(0.11)_2$ into the mantissa part, and 10000001 into the exponent part, and we have the IEEE 745 floating-point representation for 7.0:

0	10000001	1100000000000000000000
Sign	Exponent	Mantissa

The reason for using excess-127 bias in the exponent part fulfills the need to represent both positive and negative exponents. A different method, such as two's complement, could be used to represent the exponent, but excess notation facilitates faster processing. With an 8-bit exponent part, ranging from 1 to 254, the excess-127 scheme could provide an actual exponent of -127 to 128, which makes the range of floating-point numbers wide:

$$\begin{cases} \text{bias} = 1 \\ \text{exp} = \text{bias-127} \end{cases} \Rightarrow \quad \text{exp} = 1 - 127 = -126$$

$$\begin{cases} \text{bias} = 254 \\ \text{exp} = \text{bias-127} \end{cases} \Rightarrow \quad \text{exp} = 254 - 127 = 127$$

The following flowchart (Figure 3.2) shows how a decimal real number −6.25 is presented in the IEEE 754 32-bit floating-point model.

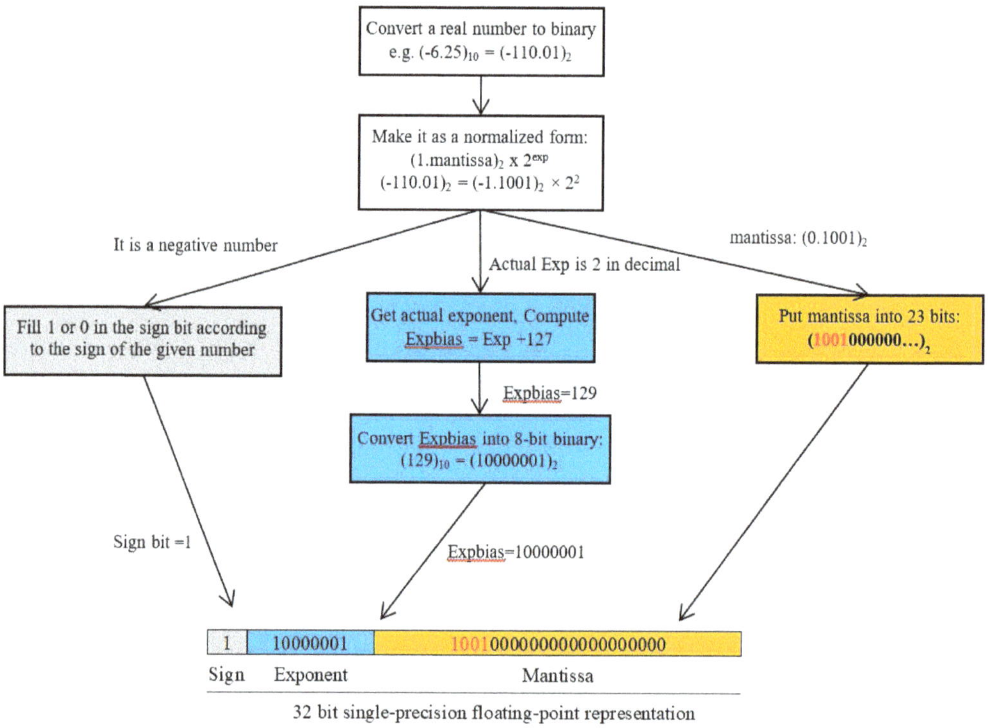

FIGURE 3.2 Flowchart Representing a Floating Number of −6.25 in the IEEE 754 Model

Example 3.1: Express 2.25 in the IEEE 754 32-bit floating-point form.

- Convert 2.25 into normalized form $(\mathbf{1.mantissa})_2 \times \mathbf{2^{exp}}$:

$$(2.25)_{10} = (10.01)_2$$

$$= (1.001)_2 \times 2^1$$

- Sign bit = 0
- The mantissa part = $(0.001)_2$
- The binary value of the exponential part is derived from exp=Expbias-127

$$\begin{cases} exp = Expbias - 127 \\ exp = 1 \quad \Rightarrow \quad Exp\ bias = 128 \end{cases}$$

$$\text{Exp bias} = (128)_{10} = (10000000)_2$$

- Thus, with $(2.25)_{10} = (1.001)_2 \times 2^1$, its IEEE 745 floating-point representation is:

0	10000000	00100000000000000000000
Sign	Exponent	Mantissa

Example 3.2: Suppose that the IEEE 754 floating-point 32-bit is

1	01111101	100 0000 0000 0000 0000 0000
Sign	Exponent	Mantissa

- Sign bit = 1, it is a negative number
- Exponent part: $(0111\ 1101)_2 = (125)_{10}$, the actual exponent value is –2:

$$\begin{cases} \text{exp} = \text{Expbias} - 127 \\ \text{Expbias} = 125 \quad \Rightarrow \quad \text{exp} = -2 \end{cases}$$

- The mantissa part is $(0.1)_2$; with an implicit leading 1, the normalized form of this floating number is:

$$(-1.1)_2 \times 2^{-2} = -1.5 \times 2^{-2} = (-0.375)_{10}$$

- Thus, this IEEE 754 floating-point representation is: $(-0.375)_{10}$

So far, we have discussed the "normalized form" **(1.mantissa)2 × 2 exp** in the IEEE 754 floating-point model, where the range of 8-bit exponent part is from 1 to 254. However, there is a serious problem: Normalized form can't represent a number 0 exactly, since the leading digit of the significand part is implicitly 1. For example, in the following representation:

0	00000000	000 0000 0000 0000 0000 0000
Sign	Exponent	Mantissa

both exponent and significand parts are zeros. The actual exponent value is $(-127)_{10}$, making the value as:

$$1.0 \times 2^{-127} = 5.87747175\text{e}{-39},$$

which is very small but not exactly a zero. Therefore, IEEE 754 defines a special case when the exponent part is 0. When the exponent part = 0, instead of the "normalized form," the numbers are switched into the "de-normalized" form:

(0.mantissa)2 × 2−126

where an implicit leading 0 (instead of 1) is used for the siginificand part, and the actual exponent is always (−126) (no excess-127 bias is applied).

Hence, if the exponent and mantissa parts are all 0s, it represents exactly the 0:

$$0.0 \times 2^{-126} = 0$$

Note: With this de-normalized form, the IEEE 754 model has two representations for 0 with the value in the sign bit being 1 and 0, respectively.

Positive zero:	0	00000000	000 0000 0000 0000 0000 0000
Negative zero:	1	00000000	000 0000 0000 0000 0000 0000

When the exponent and the significand are both zero, the quantity stored is zero. It doesn't matter what value is stored in the sign. For this reason, programmers should use caution when comparing a floating-point value to zero because negative zero does not equal positive zero.

With exponent part 0, we can also represent very small positive and negative numbers.

Example 3.3

Suppose that the IEEE 754 32-bit floating point is:

Sign	Exponent	Mantissa
1	00000000	011 0000 0000 0000 0000 0000

- Sign bit = 1; it is a negative number.
- Exponent part = $(00000000)_2$; the actual exponent value is –126, and its form is:

$$-(0.\text{mantissa})_2 \times 2^{-126}$$

- The actual mantissa is $(0.011)_2 = 1 \times 2^{-2} + 1 \times 2^{-3} = (0.375)^{10}$.
- Hence, the number is

$$-0.375 \times 2^{-126} = -4.4 \times 10^{-39},$$

which is a negative number extremely close to zero. Similary, a small positive number can be represented using the de-normalized form.

The IEEE 754 also defines another special case when the exponent part is 255.

- If the mantissa part is 0, it is + infinity $+\infty$ if the sign = 0:

Sign	Exponent	Mantissa
0	11111111	000 0000 0000 0000 0000 0000

- It is – infinity $-\infty$ if the sign = 1:

Sign	Exponent	Mantissa
1	11111111	000 0000 0000 0000 0000 0000

- If the mantissa part is not 0:

Sign	Exponent	Mantissa
0	11111111	101 0000 0000 0000 0000 0000

It is **NaN**, meaning "not a number," which is used to represent a value that is not a real number and is an error indicator.

In summary, the 32-bit floating-point representation in the IEEE 754 model has three components: 1-bit for sign, 8-bit exponent part (E), and 23-bit mantissa part (M). The floating-point value (V) is as follows:

- For 0000 0001 ≤ E ≤ 1111 1110 $(254)_{10}$, $V = (-1)^{Sign} \times 1.M \times 2^{(E-127)}$. These numbers are in the so-called normalized form. The sign-bit represents the sign of the number. The significand (1.M) is normalized with an implicit leading 1. The exponent has a bias of 127, so as to represent both positive and negative exponents. The range of the actural exponent is −126 to +127.
- For E = 0, $V = (-1)^{Sign} \times 0.M \times 2^{-126}$. These numbers are in the so-called de-normalized form. The exponent of 2^{-126} evaluates to a number very close to 0. The de-normalized form is needed to represent 0 (with F = 0 and E = 0).
- For E = 255, it represents special values, such as ±∞ (positive and negative infinity) and *NaN* (not a number).

64-bit Double-Precision Floating-Point Numbers

The IEEE 754 has a 64-bit double precision with 11-bit exponent and 52-bit mantissa. The bias is 1023. The structure of the IEEE double-precision model is shown in Figure 3.3.

Sign	Exp bias	Mantissa
1 bit	⟵ 11 bits⟶	⟵------------52 bits ------------⟶

FIGURE 3.3 The IEEE 754 64-bit Double-Precision Float-Point Representation

Chapter Summary

Integers and floats are two different kinds of numerical data. Floats are used when more precision is needed. Previously, hardware manufacturers developed their own floating-point representations. It caused data incompatibility problems due to diverse implementations. To overcome this issue, the IEEE provides a technical standard IEEE 754 to regulate the floating number formats.

There are two types of floating-point numbers defined in the IEEE 754: single-precision floating-point and double-precision floating-point numbers; 32-bit single-precision floating-point numbers consist of three parts: 1 sign bit; 8 exponent bits, using excess-127 bias (representing the exponent on a power of 2); and 23 mantissa bits. The 64-bit double precision has an 11-bit exponent using excess-1023 bias and 52-bit mantissa.

Chapter 3 Exercises

1. Translate the following scientific notation numbers into standard base-10.

   a. 2.364×10^{-3}
   b. 0.315×10^{7}
   c. 248.7695×10^{3}
   d. 31.5×10^{5}

2. Compute the IEEE 754 standard floating format for the following decimal numbers. Show each step for the process.

   a. 0.25
   b. 4.5
   c. 3.14

3. The following are floating-point numbers using the IEEE 754 representation. Compute the corresponding decimal number for each. Show each step for the process.

   a. `1 10000010 1100000 00000000 00000000`

   b. `1 10000001 1011000 00000000 00000000`

CHAPTER

4

Character Codes

Introduction

Computers were originally invented to process numbers, and we have seen how digital computers use the binary number system to represent and manipulate numeric values. Today, computers are not only used for scientific computation but also used in almost every aspect of human life. To make human–machine interaction possible, the binary data–based computer systems have to be able to process text data or character strings which are human-understandable. Thus, a variety of codes have been devised by which human-understandable characters are represented by a sequence of computer-understandable binary bit patterns. For example, Morse code was an early character coding scheme used in telecommunication. In this book, we will introduce ASCII and Unicode character codes.

Objectives

This chapter will help students:

- **gain** familiarity with the ASCII character representations.
- **understand** the meaning of character and string comparison.
- **master** the character and string operations.
- **understand** Unicode character representations.

Key Terms

ASCII: stands for American Standard Code for Information Interchange, is the numerical representation with 7 binary bits of a character. But it only covers the English writing system.

Unicode: an information technology standard for the consistent encoding, representation, and handling of text expressed in most of the world's writing systems.

non-printable character: in the ASCII table, there are 33 characters for controlling purposes. These characters, such as carriage return, line feed, etc., originated with teletype machines. There are no written symbols for these characters.

ASCII Code

ASCII, an abbreviation for American Standard Code for Information Interchange, is a character encoding standard for electronic communication. ASCII codes represent text in computers, telecommunications equipment, and other devices.

ASCII was developed from telegraph code. Its first commercial use was as a 7-bit teleprinter code promoted by Bell Data Services. The first edition of the standard was published in 1963 and used commercially for American Telephone and Telegraph's TWX (TeletypeWriter eXchange) network. It underwent a major revision in 1967 and was most recently updated in 1986.

Originally based on the English alphabet, each ASCII code is represented by a unique 7-bit value for each character. Thus, a total of 128 ($2^7 = 128$) different characters can be specified as shown by the following ASCII chart (Figure 4.1).

Compared to earlier telegraph codes, the ASCII code has added features for devices other than teleprinters.

Ninety-five of the encoded characters are **printable,** including:

- digits 0 to 9
- lowercase letters a to z
- uppercase letters A to Z
- punctuation symbols

ASCII Table

Dec	Hex	Oct	Char	Dec	Hex	Oct	Char	Dec	Hex	Oct	Char	Dec	Hex	Oct	Char
0	0	0		32	20	40	[space]	64	40	100	@	96	60	140	`
1	1	1		33	21	41	!	65	41	101	A	97	61	141	a
2	2	2		34	22	42	"	66	42	102	B	98	62	142	b
3	3	3		35	23	43	#	67	43	103	C	99	63	143	c
4	4	4		36	24	44	$	68	44	104	D	100	64	144	d
5	5	5		37	25	45	%	69	45	105	E	101	65	145	e
6	6	6		38	26	46	&	70	46	106	F	102	66	146	f
7	7	7		39	27	47	'	71	47	107	G	103	67	147	g
8	8	10		40	28	50	(	72	48	110	H	104	68	150	h
9	9	11		41	29	51	)	73	49	111	I	105	69	151	i
10	A	12		42	2A	52	*	74	4A	112	J	106	6A	152	j
11	B	13		43	2B	53	+	75	4B	113	K	107	6B	153	k
12	C	14		44	2C	54	,	76	4C	114	L	108	6C	154	l
13	D	15		45	2D	55	-	77	4D	115	M	109	6D	155	m
14	E	16		46	2E	56	.	78	4E	116	N	110	6E	156	n
15	F	17		47	2F	57	/	79	4F	117	O	111	6F	157	o
16	10	20		48	30	60	0	80	50	120	P	112	70	160	p
17	11	21		49	31	61	1	81	51	121	Q	113	71	161	q
18	12	22		50	32	62	2	82	52	122	R	114	72	162	r
19	13	23		51	33	63	3	83	53	123	S	115	73	163	s
20	14	24		52	34	64	4	84	54	124	T	116	74	164	t
21	15	25		53	35	65	5	85	55	125	U	117	75	165	u
22	16	26		54	36	66	6	86	56	126	V	118	76	166	v
23	17	27		55	37	67	7	87	57	127	W	119	77	167	w
24	18	30		56	38	70	8	88	58	130	X	120	78	170	x
25	19	31		57	39	71	9	89	59	131	Y	121	79	171	y
26	1A	32		58	3A	72	:	90	5A	132	Z	122	7A	172	z
27	1B	33		59	3B	73	;	91	5B	133	[	123	7B	173	{
28	1C	34		60	3C	74	<	92	5C	134	\	124	7C	174	|
29	1D	35		61	3D	75	=	93	5D	135	]	125	7D	175	}
30	1E	36		62	3E	76	>	94	5E	136	^	126	7E	176	~
31	1F	37		63	3F	77	?	95	5F	137	_	127	7F	177	

FIGURE 4.1 Summarized ASCII Code Chart

In addition, the original ASCII specification included 33 **non-printing** control codes that originated with teletype machines; most of these are now obsolete, although a few are still commonly used, such as:

- carriage return
- line feed
- delete and backspace
- escape
- end of line
- end of File/Stream [EOF]
- tab
- …

Character Order

Instead of exactly following the standard alphabetical order, ASCII code is ordered for more convenient sorting (i.e., alphabetization) of lists. The main differences in ASCII order are:

- All uppercase letters come before lowercase letters; for example, *Z* precedes *a*.
- Digits and many punctuation marks come before letters.

Programming on Characters

During operations (programming in C/C++ language) on characters, single characters are often involved in arithmetic and logical computation, such as comparison, addition, and subtraction. ASCII codes are used as the internal values of individual characters.

Example 4.1: What is the ASCII code of character '5'?

- According to the ASCII code chart, the code for '5' is $(53)_{10}$

Example 4.2: In C/C++ language, what is the value v of the statement: int v = '5' – '0'?

- Both '5' and '0' are character data types (char). Internally, ASCII code of the character is used for calculation
- According to the ASCII table, the codes for '5' and '0' are $(53)_{10}$ and $(48)_{10,}$ respectively
- Thus, v = 53 – 48 = 5

Example 4.3: In C/C++ language, what is the value v of the statement: int v = 'A' – 'a'?

- Both 'A' and 'a' are char data types.
- According to the ASCII table, the codes for 'A' and 'a' are $(65)_{10}$ and $(97)_{10,}$ respectively.
- Thus, v = 65 – 97 = –32.

Example 4.4: In C/C++ language, what is the c of the statement: char c = 'B'+ 38?

- 'B' is char data type, and its ASCII code is 66 in decimal.
- 38 is an integer in decimal.
- 66 + 38 =104 is in decimal.
- Since $(104)_{10}$ is stored in *c* whose data type is char (character), according to the ASCII table, *c* is the character 'h'.

Example 4.5: In C/C++ language, what is the value *b* of the statement: bool *b* = 'a'> 'Z'?

- According to the ASCII table, the codes for 'a' and 'Z' are 97 and 90 in decimal, respectively.
- 97 > 90 returns **true** in Boolean.
- Thus, *b* = **true.**

Unicode

Original ASCII codes were 7 bits long and could only encode 128 ($2^7 = 128$) characters. With the extension to 8-bit length, extended ASCII code can have 256 ($2^8 = 256$) characters in total. But it still cannot accommodate the characters of various languages around the world.

Many modern systems embrace Unicode, an information technology standard for the consistent encoding, representation, and handling of text expressed in most of the world's writing systems. The Java programming language and some operating systems, such as Windows 10, Mac OS X, Ubuntu, etc., use Unicode as their default character code.

In August 1988, a proposal for an "international/multilingual text character encoding system, tentatively called **Unicode**" was published. In October 1991, the first volume of the Unicode standard was published. This 16-bit standard *Unicode 1.0* is downward-compatible with ASCII codes and has the capacity to encode the majority of characters used in every language of the world. The codespace of *Unicode 1.0* consists of six parts, as shown in Table 4.1.

The first part is for Western alphabet codes, including English, Greek, Russian, etc.

- The lowest-numbered Unicode characters comprise the ASCII code.
- The higher codes are for the characters in other languages.

The second part is for symbolic characters, including:

- punctuation
- numbers
- pictures for control codes
- dingbats

TABLE 4.1 Outline of the *Unicode 1.0* Codespace

CHARACTER TYPES	DESCRIPTION	HEXADECIMAL RANGE
ASCII characters and other general scripts	Downward-compatible with ASCII code	0000 – 007F
	Except ASCII encodings, characters from other Latin and non-ideographic languages	0080 – 105F
Symbols	Symbolic characters including punctuation, numbers, pictures for control codes, and dingbats	2000 –278F
CJK phonetics and symbols	Punctuation marks and symbols used in characters in Chinese, Japanese, and Korean phonetic alphabets	3000 – 33FF
CJK ideographs	Ideographic Han characters used in the Chinese, Japanese, and Korean languages	4000 – 8BFF
User-defined characters	This part is reserved for the user-defined characters in their applications	E800 – FDFF
Miscellaneous	Miscellaneous glyphs, contextual or orientational variants, and width or size variants of preexisting characters.	FE00 – FFFF

(http://www.unicode.org/versions/Unicode1.0.0/ch03_6.pdf)

The third and fourth parts encode characters in Chinese, Japanese, and Korean phonetic alphabets. The fifth is reserved for special user-defined characters, and the sixth part provides variants of preexisting characters, for example, superscript or subscript glyphs of corresponding existing alphabetic characters.

Since *Unicode 2.0* (1996), Unicode was no longer restricted to 16 bits, which increased the codespace. As of 2021, the latest *Unicode 13.0* defines 143,859 characters covering 154 modern and historic scripts, as well as symbols, emoji, and non-visual control and formatting codes.

Unicode itself is a standard and can be implemented in various ways, including Unicode Transformation Formats (UTF) UTF-8, UTF-16, and UTF-32, and several others. Among them, UTF-8, the dominant encoding on the World Wide Web and on most Unix-like operating systems, uses 1 byte (8 bits) for the first 128 code points (ASCII characters) and up to 4 bytes for other characters.

Chapter Summary

Computers are not only for scientific computation, but are also used in almost every aspect of human life. To make computer systems able to process text data, ASCII codes were first devised to represent text in computers. An ASCII code is represented by a unique 7-bit value for a character. There are a total of 128 ($2^7 = 128$) different characters that can be specified. Among them, 95 characters are printable, and 33 are non-printing control codes. By using ASCII codes, characters can be compared with each other according to their numerical code values.

ASCII codes originally were based on the English writing system. In order to represent the characters in all writing systems around the world, a 16-bit Unicode was proposed in 1988, and *Unicode 1.0* was published in 1991. Unicode covers almost all characters of the writing systems in the world and is still evolving.

Chapter 4 Exercises

1. Use the ASCII table to answer the following questions:

    a. What is the octal value of the character 'A'?
    b. Suppose a char variable **v1** = **'A'+ 50**, what is the decimal integer value of v1?
    c. What is the character value of v1?
    d. Suppose an int variable **v2** = **'a' – 'A'**, what is the decimal integer value of v2?
    e. Suppose a char variable **v3** = **'B' – 9**, what is the decimal integer value of v3?
    f. What is the character value of v3?
    g. Suppose an int variable v4 = 'c' – '2', what is the decimal integer value of v4?
    h. What does 'z' > 'Z' return, True or False?

2. List the advantages of Unicode representation over ASCII code.

3. Research the Unicode representation to the hexadecimal codes for the following emojis.

a. 
Image 4.1a

b. 
Image 4.1b

c. 
Image 4.1c

Credits

Fig. 4.1: ZZT32, "Summarized ASCII Code Chart," https://commons.wikimedia.org/wiki/File:ASCII-Table.png, 2007.

IMG 4.1a: Copyright © by Emily Jäger (CC BY-SA 4.0) at https://openmoji.org/library/emoji-263A/.

IMG 4.1b: Copyright © by Laura Humpfer (CC BY-SA 4.0) at https://openmoji.org/library/emoji-2764/.

IMG 4.1c: Copyright © by Julian Grüneberg (CC BY-SA 4.0) at https://openmoji.org/library/emoji-1F44D/.

CHAPTER

5

Bitwise Operations

Introduction

Real-world information, like video, audio, image, text, and numeric data that computers can process is represented in binary format, whose smallest unit is a bit. In many situations, data can be accessed at the bit level to make operations and storage more efficient. As you will see, **bitwise operations** can be used for many processes, including setting flags, message encrypting and decrypting, and data compression and extraction.

For example, an IP address such as 192.168.1.15 has four parts. If a byte (8 bits) is used for a single part, the entire address can be packed into 32 bits of storage. The IP address can then be represented by an *unsigned int* (32-bit) in C/C++ program. This can be very handy in cases where data needs to be transmitted through a limited bandwidth network or in cases where we simply need to store data more efficiently for better memory utilization in applications. When programming mobile devices with limited memory, we often want to work at the bit level to make processing more efficient and space effective. Understanding bit operations can be useful in manipulating hardware devices and low-level programming. In this chapter, we will discuss several bitwise operations and some use cases.

Objectives

This chapter will help students:

- **understand** the motivation of bitwise operation.
- **utilize** bitwise operations: shifting, NOT, AND, OR, XOR.
- **understand** the demonstrated bitwise operation use cases.
- **develop** an application based on bitwise operations.

Key Terms

bitwise operation: used when the operand is being employed as a series of bits rather than as a whole value.

shift: an operation done on all the bits of a binary value in which they are moved by a determined number of places to either the left or right.

mask: an IP address has multiple components representing different network regions. The mask is designed to split the components from the address.

Left Shifting <<

Think of the following possibility. Suppose we want to multiply an unsigned integer by 2. We can simply **shift** all bits to the left by one position (assuming no overflow).

For example, if a 16-bit unsigned integer (*short* data type in C/C++) 41 is given:

0	0	0	0	0	0	0	0	0	0	1	0	1	0	0	1

then shift the bits by one position to the left:

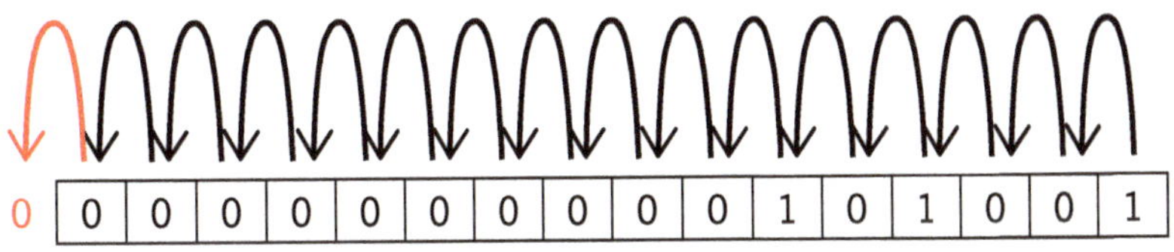

It gives us the bit pattern for an unsigned value 82, because the leftmost bit 0 (in red) is removed out of the space, and the rightmost bit is filled with a 0 after shifting:

0	0	0	0	0	0	0	0	0	0	1	0	1	0	0	1	0

Here is a snippet of C/C++ code demonstrating a left shift:

```
unsigned short a = 41;
a = a<< 1;
cout << "The variable  a is "<< a <<
endl;
```

The result of the code is:
The variable ***a*** is 82

It prints out: The variable a is 82.

The operator << is used to left shift the bits of a variable. *a* << *n* shifts the bit patterns of *a* by *n* positions resulting in the number $a \times 2^n$ assuming there is no overflow. For example:

```
unsigned short a = 5;
a = a<< 3;
cout << "The variable  a is "<<
a<< endl;
```

The result of the code is:
The variable ***a*** is 40

shifting the bits by three positions to the left:

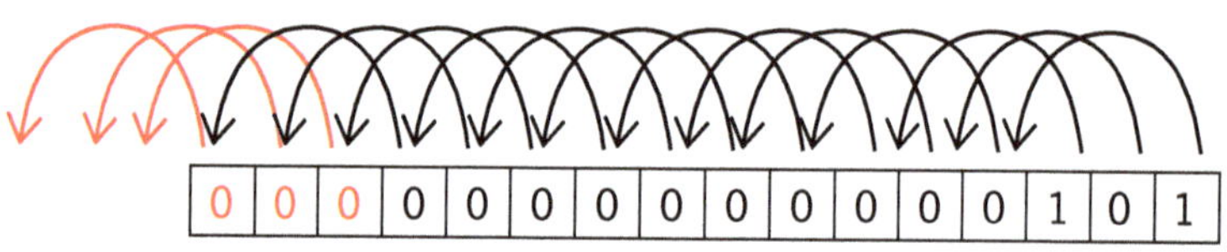

With the leftmost 3 bits 0s (in red) removed out of the space, and the rightmost 3 bits filled with 0s, the bit pattern of variable *a* is:

0	0	0	0	0	0	0	0	0	0	0	0	0	1	0	1	0	0	0

The result of the variable ***a*** is 40, which is exactly 5×2^3.

Right Shifting >>

The bit pattern of a variable can be shifted right using bit operations. For example, *x* >> *n* shifts the bit pattern of *x* by *n* positions to the right.

For example, if a 16-bit unsigned integer (short data type in C/C++) 41 is given as:

0	0	0	0	0	0	0	0	0	0	1	0	1	0	0	1

then shifting the bits by one position to the right:

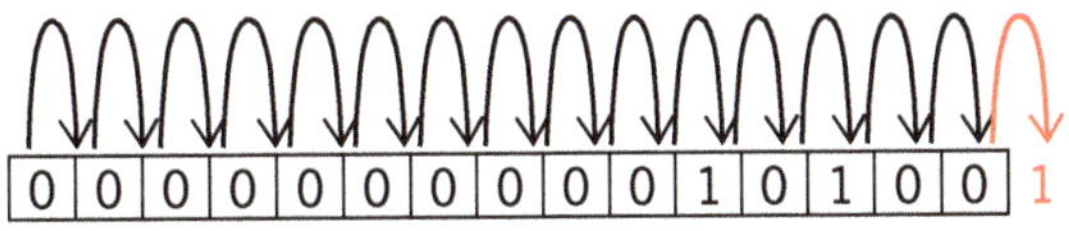

It gives us the bit pattern for *unsigned* 20, because the rightmost bit 1 is shifted out of the space, and the leftmost bit is filled with a 0 after shifting:

0	0	0	0	0	0	0	0	0	0	0	1	0	1	0	0	~~1~~

C/C++ code demonstrating the right shift:

```
unsigned short x = 141;
x = x>>2;
cout << "The variable  x is "<<
x<< endl;
```

The result of the code is:
The variable *x* is 35

shifting the bits of 141 by two positions to the right:

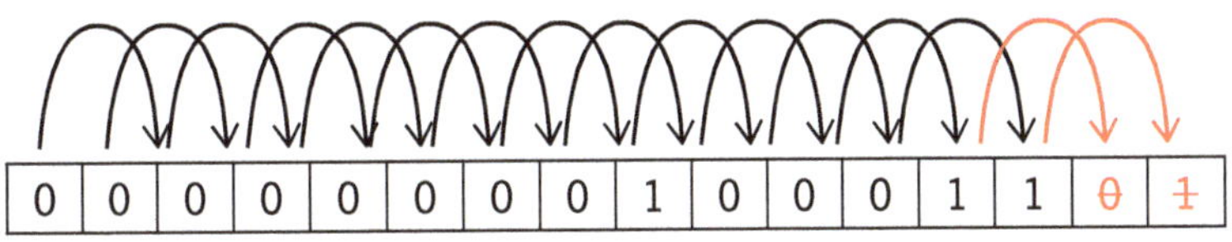

The variable *x* is 35.

0	0	0	0	0	0	0	0	0	0	1	0	0	0	1	1	~~0~~	~~1~~

Bitwise AND (&)

The bitwise AND (&) operation applies the logical AND operator on two operands in bit-by-bit fashion. Table 5.1 shows the AND operation for two single bits.

TABLE 5.1 Bitwise AND Operation

A	B	A&B
0	0	0
0	1	0
1	0	0
1	1	1

Example 5.1: 1001 AND 0101.

```
        1001
AND(&) 0101
        0001
```

The result is 0001.

In C/C++ programming, bitwise AND (&) can be used as follows:

```
#include <iostream>
using namespace std;
int main()
{
    // the bit pattern of a is 00000000 00001100
    unsigned short a = 12;
    // the bit pattern of b is 00000000 00011001
    unsigned short b = 25;
    // the bit pattern of c is 00000000 00001000
    unsigned short c = a & b;
    cout << "The variable c is: " <<  c << endl;
    return 0;
}
```

Execution result: The variable c is 8.

Bitwise OR (|)

The bitwise OR (|) operation applies the logic OR operator on two operands in bit-by-bit fashion. Table 5.2 shows the OR calculation for two single bits.

TABLE 5.2 Bitwise OR Operation

A	B	A | B
0	0	0
0	1	1
1	0	1
1	1	1

Example 5.2: 1001 OR 0101.

```
        1001
OR(|)   0101
        1101
```

The result is 1101.

Note that just applying a bit operation on a variable x does not change the value of x itself, unless we assign the value to the other variable or itself. For example, in C/C++ programming, we get a value from $x \mid 2$ and assign it back to x:

$$x = x \mid 2; \text{ or, } x \mid= 2;$$

Bitwise NEGATION (~)

The bitwise NEGATION (~) operation applies the logic NOT operator on a single operand in bit-by-bit fashion. Table 5.3 shows the operation on a single bit.

TABLE 5.3 Bitwise NEGATION Operation

A	~A
0	1
1	0

Example 5.3: NOT 1001.

```
NOT(~ ) 1001
        0110
```

The result is 0110.

Bitwise XOR (^)

Another operator of interest is the XOR (exclusive OR). The symbol of the bitwise exclusive OR operator is ^. Table 5.4 shows the XOR calculation for two single bits.

TABLE 5.4 Bitwise XOR Operation

A	B	A ^ B
0	0	0
0	1	1
1	0	1
1	1	0

Example 5.4: 0001 XOR 0101.

```
              0001
XOR(^)        0101
              0100
```

It results in 0100.

The following is an XOR operation example in C/C++ programing: there are four char variables: *x*, *e*, *d*, and *key*.

```
#include <iostream>
using namespace std;
int main()
{
        char x='A';
        char key='b';
        char e= x ^ key;   // applying XOR on x and key
        char d= e ^ key; // applying XOR on e and key
}
```

What are the characters in variables of *e* and *d*, respectively?

- According to the ASCII code Table 5.4, the binary patterns of character 'A' and '*b*' are

  0100 0001 and 0110 0010, respectively

- By applying XOR on *x* and *key* (*e*= *x* ^ *key*):

	0100 0001	(*x: ASCII code of* 'A')
^	0110 0010	(*key: ASCII code of* '*b*')
	0010 0011	(*e: ASCII code of* '#')

  The bit pattern of *e* is 0010 0011(0×23). Check the ASCII table; the character of *e* is '#''.

- Applying XOR again on *e* and *key* (*d*= *e* ^ *key*):

	0010 0011	(*e: ASCII code of* '#')
^	0110 0010	(*key: ASCII code of* '*b*')
	0100 0001	(*d: ASCII code of* 'A')

The binary value of *d* is 0100 0001 (0×41), which is the character 'A' according to the ASCII table. This is the same value that the variable *x* was assigned in the program. Therefore, the characters assigned to the variables *e* and *d* are '#' and 'A', respectively.

From this example, we see that XOR has an interesting property: it can be used for simple data encrypting and decrypting operations. In our example, after encrypting the original data 'A' with a key '*b*', the encrypted data becomes a meaningless symbol '#', which seemingly has no obvious relationship with the original character 'A'. If you have the key '*b*' and know how to decrypt the data (by XOR it again), the original data can be restored. Note the *key* can be any data as long as it is consistent when encrypting and decrypting. You may use this "trick" to develop a simple encrypting and decrypting tool for fun.

Bit Operation in C/C++ Programming

The following C++ program uses all the bitwise operators mentioned earlier. Please read and understand it.

```
#include <iostream>
using namespace std;
int main()
{
    unsigned int a = 60; /* 60 = 0011 1100 */
    unsigned int b = 13; /* 13 = 0000 1101 */
    int c = 0;

    c = a & b;        /* 12 = 0000 1100 */
    cout << "Line 1 - Value of c is "<< c <<endl;

    c = a | b;       /* 61 = 0011 1101 */
    cout << "Line 2 - Value of c is "<< c <<endl;

    c = a ^ b;        /* 49 = 0011 0001 */
    cout << "Line 3 - Value of c is "<< c <<endl;

    c = ~a;           /*-61 = 1100 0011 */
    cout << "Line 4 - Value of c is "<< c <<endl;

    c = a << 2;      /* 240 = 1111 0000 */
    cout << "Line 5 - Value of c is "<< c <<endl;

    c = a >> 2;      /* 15 = 0000 1111 */
    cout << "Line 6 - Value of c is "<< c <<endl;

    return 0;
}
```

When you compile and execute the above program, it produces the following results:

Line 1—Value of c is 12.
Line 2—Value of c is 61.
Line 3—Value of c is 49.
Line 4—Value of c is −61.
Line 5—Value of c is 240.
Line 6—Value of c is 15.

Table 5.5 provides the summarized explanation. Please note:

- The data type used here, unsigned int, has 32 bits. However, for simplicity, only the low 8 bits are shown because the values in this example are small. All other higher bits are zeros.
- The bit pattern in line 4: 1100 0011 is in the form of two's complement. The system is able to adjust it and get −61 (for the details, check the previous two's complement section).

TABLE 5.5 Explanation of Above Program (a = 60, b = 13, their bit patterns are 00111100 and 00001101, respectively)

OPERATOR	DESCRIPTION	CODE EXAMPLE
<<	Binary Left Shift Operator. The left operand's value is moved left by the number of bits specified by the right operand.	a << 2 = 240 i.e., 1111 0000
>>	Binary Right Shift Operator. The left operand's value is moved right by the number of bits specified by the right operand.	a >> 2 = 15 i.e., 0000 1111
&	Binary AND Operator copies a bit to the result if it exists in both operands.	(a & b) = 12, i.e., 0000 1100
|	Binary OR Operator copies a bit if it exists in either operand.	(a | b) = 61, i.e., 0011 1101
~	Binary One's Complement Operator is unary and has the effect of "flipping" bits.	(~a) = −61, i.e., 1100 0011
^	Binary XOR Operator copies the bit if it is set in one operand but not both.	(a ^ b) = 49, i.e., 0011 0001

A Use Case of Bit Operations

In many situations. we need to find a specific bit or a group of bits from the given data. For example, an IP address 192.168.2.16 has four parts and can be packed into a 32-bit unsigned integer whose bit patterns are:

1 1 0 0 0 0 0 0	1 0 1 0 1 0 0 0	0 0 0 0 0 0 1 0	0 0 0 1 0 0 0 0
192	168	2	16

where each byte (8 bits) of this 32-bit unsigned integer stores one part of this IP address. We may want to extract some part(s), such as the high-order 8 bits, "192" from this address representation. This can be done by using an appropriate **mask** and applying a bitwise AND operation with shifting. To extract the number in the high-order 8 bits for this example, the steps are:

- Define a mask, a 32-bit unsigned integer whose bit pattern is:

1 1 1 1 1 1 1 1	0 0 0 0 0 0 0 0	0 0 0 0 0 0 0 0	0 0 0 0 0 0 0 0
255	0	0	0

  This mask value has all *1*s in the high-order 8 bits, and the rest are 0s. The hexadecimal value of this mask is 0x FF 00 00 00.
- Take bitwise AND operator (&) as follows:

	11000000 10101000 00000010 00010000	(IP address)
&	11111111 00000000 00000000 00000000	(Mask)
	11000000 00000000 00000000 00000000	(Result)

  The result of this step preserves the data in the high-order 8 bits of the IP address and masks the low-order 24 bits with 0s.
- Right shift (>>) the masked result by 24 steps:

  11000000 00000000 00000000 00000000 >> 24

- The bit patterns of the shifted result are:

  00000000 00000000 00000000 11000000

  where the value of the high-order 8 bits of the IP address is shifted to the rightmost 8 bits in the 32-bit unsigned integer.
- Converting the shifted result into decimal, we get the desired result of *192*.

If we implement the process in C/C++ language, the code and the execution result is:

```
#include <iostream>
using namespace std;
int main()
{
   // Hex format of the 32-bit IP address
192.168.2.16
   unsigned int IP = 0xC0A80210;
   // Hex format of the 32-bit mask 255.0.0.0
   unsigned int mask = 0xFF000000;
   unsigned int Last_part_ip = (IP & mask) >> 24;
   cout << "The last part of the given IP
address is: " <<  Last_part_ip << endl;
   return 0;
}
```

```
$g++ -o main
*.cpp
$main
The last part
of the given
IP address is:
192
```

In a similar way, you can extract out any part of an IP address.

Chapter Summary

Bitwise operations can be used for many applications, such as setting flags, message encrypting and decrypting, data compression and extraction, and IP address masking.

Several logical bitwise operations are introduced:

- Shift left
- Shift right
- Bitwise AND
- Bitwise OR
- Bitwise NOT
- Bitwise XOR.

Chapter 5 Exercises

1. Performing the following bit operations:

   a. 0101 1100 AND 1111 0000
   b. 0101 1100 OR 0011 0101
   c. NOT 1010 1100
   d. 1001 1101 XOR 0111 1000

2. Get the results of the following bit operations for a binary number V = 1011 1000:

   a. V >> 1
   b. V << 2
   c. (V >> 4) AND 0x0F
   d. V XOR 0x1A XOR 0x1A

3. Show the decimal results of bit shifting operations for the following problems. First translate the number into binary, perform the bit shifting and translate the result back to decimal.

   a. 192 >> 1
   b. 192 >> 2
   c. 192 >> 3
   d. 192 >> 4
   e. 8 << 1
   f. 8 << 2
   g. 8 << 3
   h. 8 << 4

4. In a TCP/IP package, the first 16 bits encode three pieces of information: version, length and type of service. The information is arranged as follows:

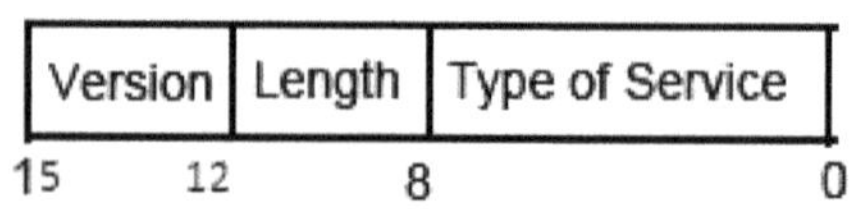

If you have this 2-byte value in a variable T, use bit operations to extract the **version** and **length** info into variables V and L, respectively.

5. Use bit shifting and logical operations to extract the high-order 8 bits and low-order 8 bits of the following word V into unsigned char variables H and L, respectively.

V:	1	1	1	1	0	0	0	0	0	0	0	0	1	1	1	1

CHAPTER

6

Boolean Function

Introduction

This chapter covers Boolean operations, truth tables, Boolean expressions, gates, logic diagrams, and basic simplification of Boolean expressions. The basic logical AND, OR, and NOT operations are used to implement complex processes used in digital circuits. The conversions among Boolean expressions, truth tables, and logic diagrams are discussed. According to basic identities of Boolean algebra, Boolean expressions can be simplified.

Objectives

This chapter will help students:

- **understand** Boolean operations, truth tables, Boolean expressions, gates, logic diagrams, and basic simplification of Boolean expressions.
- **master** Boolean operations, truth tables, Boolean expressions, gates, and logic diagrams.
- **implement** basic identities of Boolean algebra and basic simplification of Boolean expressions.

Key Terms

AND operator: returns the Boolean value true if both operands are true and returns false otherwise.

OR operator: returns the Boolean value true if either or both operands are true and returns false otherwise.

NOT operator: reverses the true/false outcome of the expression that immediately follows.

Boolean function: a function whose arguments and result assume values from a two-element set (usually {true, false} or {0,1}).

truth table: a table used to illustrate and determine the truth value of propositions and the validity of their resulting argument.

logic gate: acts as a basic building block for digital circuits.

logic diagram: graphical illustration that shows how different types of gates are connected to form a digital circuit to perform a particular task.

sum-of-product: expresses the combinational logic (a Boolean algebra expression) as two or more products summed together.

Boolean Operations

A Boolean expression consists of one or multiple variables and operators and produces a result of 0 or 1. Boolean expressions are often applied in programming languages to quantitatively evaluate different logic operations.

AND, *OR*, and *NOT* are three common logical operators. As shown in Tables 6.1 to 6.3, the related operations of these three logical operators are given in truth tables. A truth table shows all possible combinations of inputs and the output results for these operations.

The logical operator AND is typically represented as xy or $x \cdot y$, which is often referred to as a Boolean product and read as "x and y." As shown in Table 6.1, the AND operation gives a value of true (marked as 1) only if all inputs are true. Otherwise, the AND operation result is false.

The logical operator OR is typically represented as $x + y$, which is often referred to as a *Boolean sum* and read as "x or y." As shown in Table 6.2, the OR operation has a result of false (marked as 0) only when the two inputs are false. Otherwise, the OR operation result is true.

The logical operator NOT is typically represented as $\overline{x}$ or x', which is read as "NOT *x*." As shown in Table 6.3, the NOT operation flips the input value. If the input is true (marked as 1), the output is false (marked as 0). Similarly, if the input is false, the output is true.

TABLE 6.1 The Truth Table for AND

INPUTS		OUTPUTS
x	y	xy
0	0	0
0	1	0
1	0	0
1	1	1

TABLE 6.2 The Truth Table for OR

INPUTS		OUTPUTS
x	y	$x + y$
0	0	0
0	1	1
1	0	1
1	1	1

TABLE 6.3 The Truth Table for NOT

INPUTS	OUTPUTS
x	$\overline{x}$
0	1
1	0

A Boolean expression may involve one or multiple binary variables and logical operators. Various Boolean expressions can be composed by the different combinations of binary variables and logical operators. For example, a Boolean function $F(x, y, z) = \overline{x} + y\,\overline{z}$ as shown in Table 6.4 is represented by a Boolean expression involving three Boolean variables x, y, and z and three logical operators OR, NOT, and AND. The three logical operators in the Boolean function $F(x, y, z) = \overline{x} + y\,\overline{z}$ have the corresponding order of operations, NOT > AND > OR. NOT has the highest priority. For the calculation of the function F, x and z are first negated, then the AND of y and $\overline{z}$ is performed, and finally the AND result is OR with $\overline{x}$.

TABLE 6.4 The Truth Table for $F(x, y, z) = \overline{x} + y\,\overline{z}$

INPUTS			INTERMEDIATE			OUTPUTS
x	y	z	$\overline{x}$	$\overline{z}$	$y\,\overline{z}$	$\overline{x} + y\,\overline{z} = F$
0	0	0	1	1	0	1
0	0	1	1	0	0	1
0	1	0	1	1	1	1
0	1	1	1	0	0	1
1	0	0	0	1	0	0
1	0	1	0	0	0	0
1	1	0	0	1	1	1
1	1	1	0	0	0	0

A truth table is often applied to break down a complex function. The components of the (complex) function are represented in the truth table. As shown in Table 6.4, the function is finally evaluated by combining the result of each component. The intermediate steps clearly illustrate the entire process of reaching the complete evaluation of the function.

There are four more examples as follows.

Example 6.1

The truth table for $F(x, y) = \overline{xy}$ is given in Table 6.5.

TABLE 6.5 The Truth Table for $F(x, y) = \overline{xy}$

INPUTS		INTERMEDIATE	OUTPUTS
x	y	xy	$\overline{xy}$
0	0	0	1
0	1	0	1
1	0	0	1
1	1	1	0

Example 6.2

The truth table for $F(x, y, z) = yz + \overline{xy}z$ is given in Table 6.6.

TABLE 6.6 The Truth Table for $F(x, y, z) = yz + \overline{xy}z$

INPUTS			INTERMEDIATE				OUTPUTS
x	y	z	yz	xy	$\overline{xy}$	$\overline{xy}z$	$yz + \overline{xy}z$
0	0	0	0	0	1	0	0
0	0	1	0	0	1	1	1
0	1	0	0	0	1	0	0
0	1	1	1	0	1	1	1
1	0	0	0	0	1	0	0
1	0	1	0	0	1	1	1
1	1	0	0	1	0	0	0
1	1	1	1	1	0	0	1

Example 6.3

The truth table for $F(x, y, z) = x(\overline{y} + z) + xyz$ is given in Table 6.7.

TABLE 6.7 The Truth Table for $F(x, y, z) = x(\overline{y} + z) + xyz$

INPUTS			INTERMEDIATE				OUTPUTS
x	y	z	$\overline{y}$	$\overline{y} + z$	$x(\overline{y} + z)$	xyz	$x(\overline{y} + z) + xyz$
0	0	0	1	1	0	0	0
0	0	1	1	1	0	0	0
0	1	0	0	0	0	0	0
0	1	1	0	1	0	0	0
1	0	0	1	1	1	0	1
1	0	1	1	1	1	0	1
1	1	0	0	0	0	0	0
1	1	1	0	1	1	1	1

Example 6.4

The truth table for $F(x, y) = (x + y)(\overline{x} + y)$ is given in Table 6.8.

TABLE 6.8 The Truth Table for $F(x, y) = (x + y)(\overline{x} + y)$

INPUTS		INTERMEDIATE			OUTPUTS
x	y	$\overline{x}$	$x + y$	$\overline{x} + \overline{y}$	$(x + y)(\overline{x} + y)$
0	0	1	0	1	0
0	1	1	1	1	1
1	0	0	1	0	0
1	1	0	1	1	1

Boolean Expression

A Boolean function can be represented in either a truth table or a Boolean expression. A Boolean function can be represented by many different logically equivalent expressions. If two expressions can be represented by the same truth table, they are considered logically equivalent (refer to Example 6.5).

Example 6.5

Suppose $F(w, x, y, z) = \overline{w}x + x\overline{y}z$. According to idempotent law, F can also be expressed as $F(w, x, y, z) = \overline{w}x + \overline{w}x + x\overline{y}z$. Additionally, F can also be expressed as $F(w, x, y, z) = x(\overline{w} + \overline{y}z)$ using the distributive law.

The sum-of-products form and the product-of-sums form are two of the most common forms used in Boolean functions.

Sum-of-products form: The expression of a Boolean function must be a collection of variables connected by AND operations. For instance, the function $F1(x, y, z) = \overline{x}y + \overline{y}z + x\overline{y}\overline{z}$ is in sum-of-products form. The function $F2(x, y, z) = \overline{x}y + \overline{y}(z + x\overline{z})$ is not in sum-of-products form. The distributive law can be applied to distribute the $\overline{y}$ variable in F2, resulting in the expression $\overline{x}y + \overline{y}z + x\overline{y}\overline{z}$ (*same as* $F1(x, y, z)$) in sum-of-products form.

Product-of-sums form: The expression of Boolean function must consist of variables connected by OR operations. For instance, the function $F1(x, y, z) = (x + \overline{y})(\overline{x} + z)(y + \overline{z})$ is in product-of-sums form.

Since any Boolean expression can be represented as a truth table, any Boolean expression can be represented in sum-of-products form. Which form is better? It depends on the usage. The truth value of an expression is easy to evaluate in the product-of-sums form.

An expression in sum-of-products form may not be in simplest form. This means there may be an easier way to calculate the result of the expression. Both sum-of-products and product-of-sums forms of a Boolean function can be mutually converted. The conversion can reduce the expression in sum-of-products or product-of-sums to the minimum number of terms. The minimum number of terms is conducive to the implementation of electrical circuits.

Example 6.6

Consider a simple majority function. When given three inputs, the majority function outputs a 0 if more than half of its inputs are 0, and a 1 if more than half of its inputs are 1. Table 6.9 depicts the truth table for the majority function over three variables.

TABLE 6.9 Truth Table Representation for the Majority Function

X	Y	Z	F
0	0	0	0
0	1	0	0
0	0	1	0
0	1	1	1
1	0	0	0
1	1	0	1
1	0	1	1
1	1	1	1

When $x = 0$, $y = 0$, and $z = 0$, the function must output a 0. The product term that satisfies this is xyz. When $x = 1$, $y = 0$, and $z = 1$, the function must output a 1. The product term that satisfies this is $x\bar{y}z$. According to the truth table, a product term of the input variables x, y, and z can be easily generated.

Boolean Algebra

As a branch of algebra generally used in mathematical logic, Boolean algebra only uses two values usually denoted as 1 and 0 to represent the logic meanings such as true and false, on and off. It was first introduced by George Boole in *The Mathematical Analysis of Logic* (1847). Boolean logic is the foundation for the development of digital electronics and is widely used in modern programming languages. Computers consisting of CPU, graphics card, and motherboard are built by the collections of electronic components. These electronic components have two working statuses, either on or off. Boolean algebra can accurately represent the working status of each electronic component. Generally, 0 and 1 are used to represent false and true, respectively.

Boolean Identities

Given a regular algebra expression, it may not be in its simplest form. For instance, a regular algebraic expression $5x + 11x$ can be further merged to obtain $16x$. Therefore, a regular algebra expression may be simplified by expressing the result using fewer or simpler terms. Similarly, a Boolean expression may not be in its simplest form, which means partial terms of this Boolean expression can be merged through certain operations. Table 6.10 shows the basic identities for Boolean variables and expressions. Except for the double complement law, each relationship has both an AND form and an OR form, which is known as the *duality principle*.

TABLE 6.10 Basic Identities of Boolean Algebra

IDENTITY NAME	AND FORM	OR FORM
Identity law	$1x = x$	$0 + x = x$
Null (or Dominance) law	$0x = 0$	$1 + x = 1$
Idempotent law	$xx = x$	$x + x = x$
Inverse law	$x\overline{x} = 0$	$x + \overline{x} = 1$
Commutative law	$xy = yx$	$x + y = y + x$
Associative law	$(xy)z = x(yz)$	$(x + y) + z = x + (y + z)$
Distributive law	$x + yz = (x + y)(x + z)$	$x(y + z) = xy + xz$
Absorption law	$x(x + y) = x$	$x + xy = x$
De Morgan's law	$(\overline{xy}) = \overline{x} + \overline{y}$	$(\overline{x + y}) = \overline{x}\overline{y}$
Double Complement law	$\overline{\overline{x}} = x$	

Table 6.11 shows a truth table demonstrating the AND form of De Morgan's law $(\overline{xy}) = \overline{x} + \overline{y}$, and proves that $(\overline{xy}) \neq \overline{x}\,\overline{y}$.

TABLE 6.11 Truth Tables for the AND Form of De Morgan's Law

x	y	(xy)	$(\overline{xy})$	$\bar{x}$	$\bar{x}$	$\bar{x}\,\bar{y}$	$\bar{x}+\bar{y}$
0	0	0	1	1	1	1	1
0	1	0	1	1	0	0	1
1	0	0	1	0	1	0	1
1	1	1	0	0	0	0	0

Simplification of Boolean Expressions

According to the above discussion of algebraic identities, the algebraic expression $(5x + 4y - 2x + 2y)$ *can be further reduced to* the simplest forms $(3x + 6y)$. The Boolean identities can be used to simplify Boolean expressions in a similar fashion.

Example 6.7

Suppose there is a function $F(x, y) = xy + xy$.

According to the OR form of the idempotent law, the expression xy is treated as a Boolean variable, and then the original expression $xy + xy$ is simplified to xy. Therefore, $F(x, y) = xy + xy = xy$.

Example 6.8

Given a function $F(x, y, z) = x\bar{y}\bar{z} + \bar{x}\bar{y}\bar{z} + \bar{x}z$, it is simplified as follows:

$$F(x, y, z) = x\bar{y}\bar{z} + \bar{x}\bar{y}\bar{z} + \bar{x}z$$

$$= (x + \bar{x})\,\bar{y}\bar{z} + \bar{x}z \text{ (Distributive)}$$

$$= (1)\,\bar{y}\bar{z} + \bar{x}z \text{ (Inverse)}$$

$$= \bar{y}\bar{z} + \bar{x}z \text{ (Identity)}$$

The simplification is straightforward in the above two examples. However, identity extension is often used in the simplification process to obtain the simplest form. For example, 1 is extended to $x + \bar{x}$ in the intermediate simplification process.

Example 6.9

Given the function $F(x, y, z) = x\bar{y} + \bar{x}\bar{z} + \bar{y}\bar{z}$, it can be simplified as follows:

$= x\bar{y} + \bar{x}\bar{z} + \bar{y}\bar{z}(1)$ (Identity)

$= x\bar{y} + \bar{x}\bar{z} + \bar{y}\bar{z}(x + \bar{x})$ (Inverse)

$= x\bar{y} + \bar{x}\bar{z} + (\bar{y}\bar{z})x + (\bar{y}\bar{z})\bar{x}$ (Distributive)

$= x\bar{y} + \bar{x}\bar{z} + x(\bar{y}\bar{z}) + \bar{x}(\bar{z}\bar{y})$ (Commutative)

$= x\bar{y} + \bar{x}\bar{z} + (x\bar{y})\bar{z} + (\bar{x}\bar{z})\bar{y}$ (Associative)

$= x\bar{y} + (x\bar{y})\bar{z} + \bar{x}\bar{z} + (\bar{x}\bar{z})\bar{y}$ (Commutative)

$= x\bar{y}(1 + \bar{z}) + \bar{x}\bar{z}(1 + \bar{y})$ (Distributive)

$= x\bar{y}(1) + \bar{x}\bar{z}(1)$ (Null)

$= x\bar{y} + \bar{x}\bar{z}$ (Identity)

Example 6.9 is commonly known as the *Consensus Theorem*. Which cases are suitable for identity extension in the simplification process? Unfortunately, there is no rule for identity extension. The use of identity extension mainly depends on experience.

Identities can be used to prove Boolean equalities. The proof of the function $(x + \bar{y})(x + y) = x$ is given as follows:

$(x + \bar{y})(x + y) = xx + xy + \bar{y}x + \bar{y}y$ Distributive law

$= xx + xy + \bar{y}x + 0$ Inverse law

$= x + xy + \bar{y}x + 0$ Idempotent law

$= x + xy + \bar{y}x$ Identity law

$= x + x(y + \bar{y})$ Distributive law (and Commutative law)

$= x + x(1)$ Inverse law

$= x + x$ Identity law

$= x$ Idempotent law

The truth tables of the function $(x + \bar{y})(x + y)$ can also be used to prove the equality of two Boolean expressions $(x + \bar{y})(x + y)$ and x. If the truth tables are identical, the expressions are equal. According to the truth table shown in Table 6.12, the value of x is exactly equal to the value of the function $(x + \bar{y})(x + y)$, when x and y are in different combinations of 0 and 1 values.

TABLE 6.12 The Truth Table of the Function $(x + \bar{y})(x + y)$

x	y	$\bar{y}$	$x + \bar{y}$	$x + y$	$(x + \bar{y})(x + y)$
0	0	1	1	0	0
1	0	1	1	1	1
0	1	0	0	1	0
1	1	0	1	1	1

Complements

As shown in Example 6.7, the Boolean identities can be applied to Boolean expressions, as well as the Boolean operators. For example, *xy is first treated* as a Boolean variable, and then the idempotent law is applied. The *complement* of the expression is involved when a common Boolean operator is applied to various complex Boolean expressions. Compared with the implementation of a function itself, sometimes it is easy to implement the complement of the function. After the implementation of the complement, the result of the complement needs to be inverted to obtain the result of the original function as the final result. In this case, complements are conducive to obtaining the results of the original functions.

De Morgan's law is applied to obtain the complement of a Boolean function, such as the OR form $(\overline{x + y}) = \overline{xy}$. The OR form can be easily extended to three or more variables as follows:

Given the function:

$$F(w, x, y, z) = (w + x + y + z), \text{ then } (w, x, y, z) = (\overline{w + x + y + z})$$

Let $u = (w + y)$ and $v = (y + z)$. Then,

$$\overline{F}(w, x, y, z) = \overline{(u + v)} = \overline{uv}$$

Now, the following equation is obtained after applying De Morgan's law again.

$$\begin{aligned} \overline{F}(w, x, y, z) &= \overline{(u + v)} = \overline{uv} \\ &= \overline{(w + x)}\,\overline{(y + z)} \\ &= \overline{wxyz} \end{aligned}$$

Therefore, if $F(w, x, y, z) = (w + x + y + z)$, then $\overline{F}(w, x, y, z) = \overline{w}\overline{x}\overline{y}\overline{z}$. Similarly, $\overline{(wxyz)} = \overline{w} + \overline{x} + \overline{y} + \overline{z}$ can be obtained after applying De Morgan's law.

In order to obtain the complement of a Boolean expression, each variable is simply replaced by its complement (x is replaced by $\overline{x}$), and ANDs and ORs are interchanged. For example, according to De Morgan's law, the complement of $F(x, y, z) = \overline{x} + \overline{y}\,z$ is $\overline{F}(x, y, z) = x(y + \overline{z})$. The shaded results shown in Table 6.13 confirm $\overline{\overline{x} + \overline{y}\,z} = x(y + \overline{z})$.

TABLE 6.13 Truth Table Representation for a Function and Its Complement

x	y	z	$\overline{x}$	$\overline{y}$	$\overline{z}$	$\overline{y}\,z$	$\overline{x} + \overline{y}z$	$y + \overline{z}$	$\overline{\overline{x} + \overline{y}\,z}$	$x(y + \overline{z})$
0	0	0	1	1	1	0	1	1	0	0
0	1	0	1	0	1	0	1	1	0	0
0	0	1	1	1	0	1	1	0	0	0
0	1	1	1	0	0	0	1	1	0	0
1	0	0	0	1	1	0	0	1	1	1
1	1	0	0	0	1	0	0	1	1	1
1	0	1	0	1	0	1	1	0	0	0
1	1	1	0	0	0	0	0	1	1	1

Logic Gates

AND, OR, and NOT as three basic logical operators are used to compose different Boolean expressions. Large assemblies of logic gates (as basic logic elements for digital functions) are often packaged in integrated circuits consisting of digital circuits to implement arithmetic operations. Boolean expressions can be represented by logic gates to compose complex devices. A logic gate is an electronic device used to implement a Boolean function. It processes a logic operation on one or more binary inputs to produce a single binary output.

Symbols for Logic Gates

Figure 6.1 visualizes three basic logic gates that correspond to AND, OR, and NOT operators. The circle at the output of the NOT gate typically represents the complement operation.

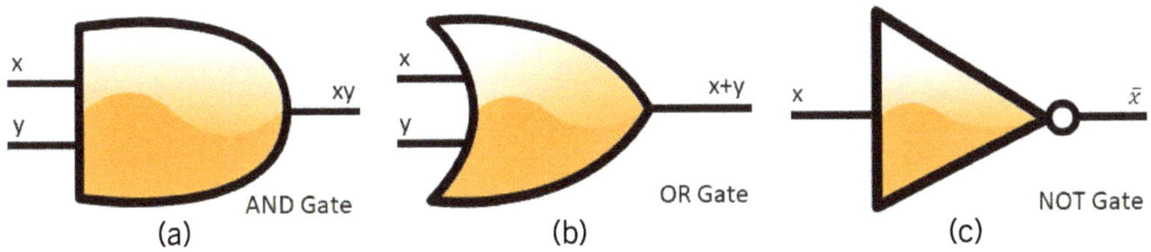

FIGURE 6.1A-C Three Basic Gates

The exclusive-OR (XOR) gate is represented by the Boolean expression: $x \oplus y$. XOR is false if both input values are equal and true otherwise. The logical XNOR operation is represented by ⊙. That is a dot surrounded by a circle. The function of XNOR is the logical complement of the XOR gate. The truth table and the logic diagrams of XOR and XNOR are shown in Table 6.14 and Figure 6.2, respectively.

TABLE 6.14 The Truth Table for XOR and XNOR

X	Y	X XOR Y	X XNOR Y
0	0	0	1
0	1	1	0
1	0	1	0

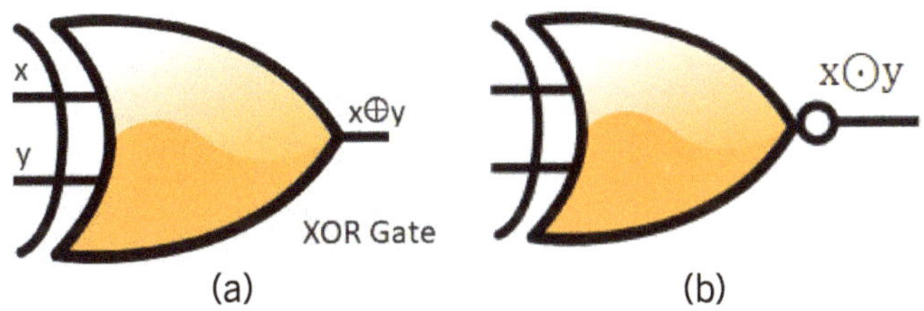

FIGURE 6.2A-B The Logic Symbols for XOR Gate and XNOR Gate

Universal Gates

NAND and NOR as common gates produce complementary output to AND and OR, respectively. As shown in Tables 6.15 and 6.16 and Figures 6.3 and 6.4, the logic diagrams for NAND and NOR along with the truth tables have two different gate representations (straightforward representation and converted representation using De Morgan's law).

TABLE 6.15 The Truth Table for NAND

X	*Y*	*X* NAND *Y*
0	0	1
0	1	1
1	0	1
1	1	0

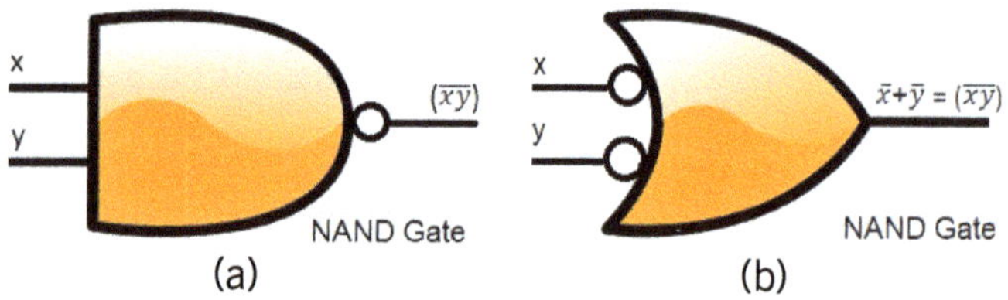

FIGURE 6.3A-B The Logic Symbols for NAND Gate

TABLE 6.16 The Truth Table for NOR

X	*Y*	*X* NOR *Y*
0	0	1
0	1	0
1	0	0
1	1	0

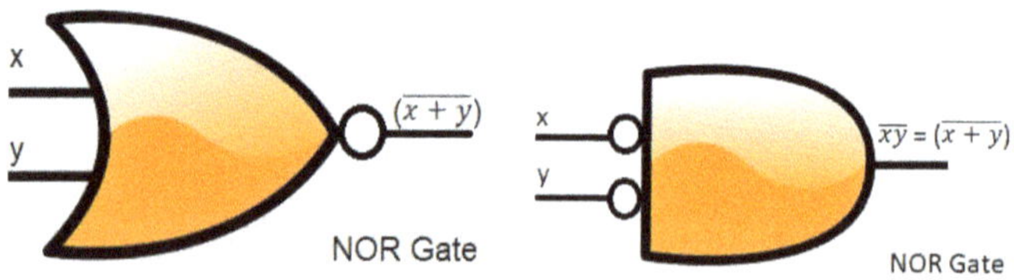

FIGURE 6.4 The Logic Symbols for NOR Gate

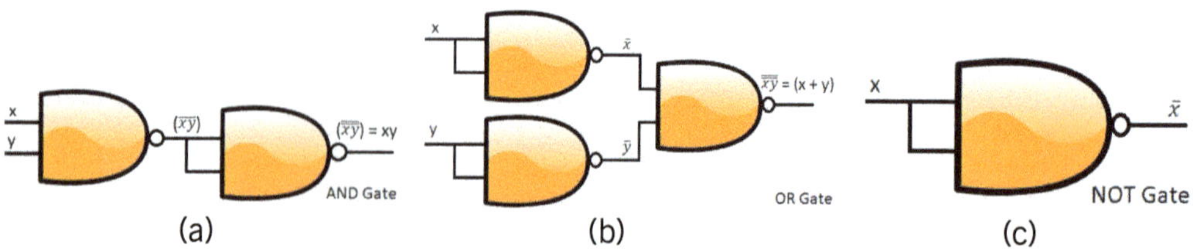

FIGURE 6.5A-C AND, OR, and NOT Gates Constructed Using Only NAND Gates

The NAND gate as a *universal gate* can be used to construct any other gates used in electronic circuits. According to Figure 6.5, an AND gate, an OR gate, and a NOT gate can be represented by only using NAND gates.

Is there any benefit from the universal gate? Definitely, yes. You may ask why the NAND gates are often used to construct the AND, OR, and NOT gates rather than directly using the AND, OR, and NOT gates. There are two main reasons for using only NAND gates in building circuits. First, the cost of building NAND gates is lower than the cost of building other gates. Second, using the same type of gate to build complex integrated circuits is much easier than using different types of gates.

According to the duality principle, any circuit can also be built by using only NOR gates (similar as only using NAND gates). NAND and NOR can be applied to the expression implementations of sum-of-products form and product-of-sums form, respectively.

Multiple Input Gates

In the above discussion, all gates only have two inputs. Actually, a gate is allowed to have multiple inputs and outputs. As shown in Figures 6.6 and 6.7, the gates have three inputs.

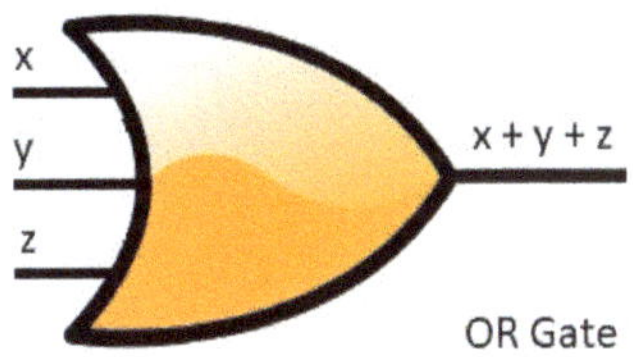

FIGURE 6.6 A Three-Input x, y, and z OR Gate Representing $x + y + z$

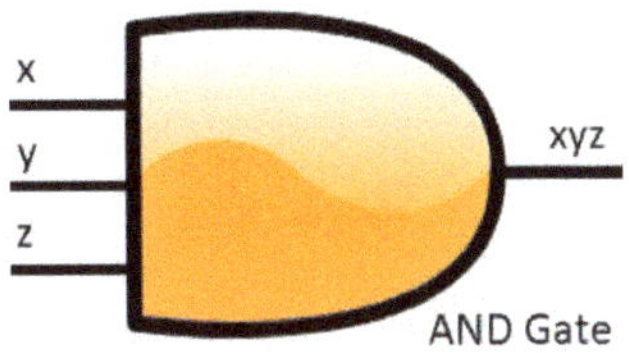

FIGURE 6.7 A Three-Input x, y, and z AND Gate Representing xyz

As shown in Figure 6.8, the AND gate has two inputs and two outputs Q and $\overline{Q}$. Q always represents the actual output.

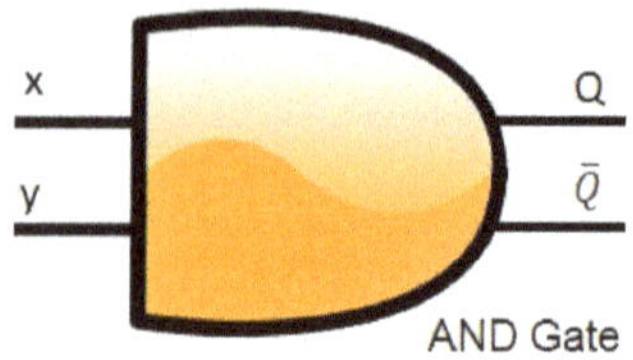

FIGURE 6.8 An AND Gate with Two Inputs and Two Outputs

Logic Diagram

How do we implement Boolean functions on digital circuits? A simple logic gate can represent a simple Boolean operation (such as AND or OR). The combinations of AND, OR, and NOT gates can represent complex Boolean expressions. Therefore, the physical implementation of any expression can be represented in a logic diagram using various combinations of gates. As shown in Figure 6.9, the function $F(x, y, z) = \bar{x} + y\bar{z}$ is represented in a logic diagram.

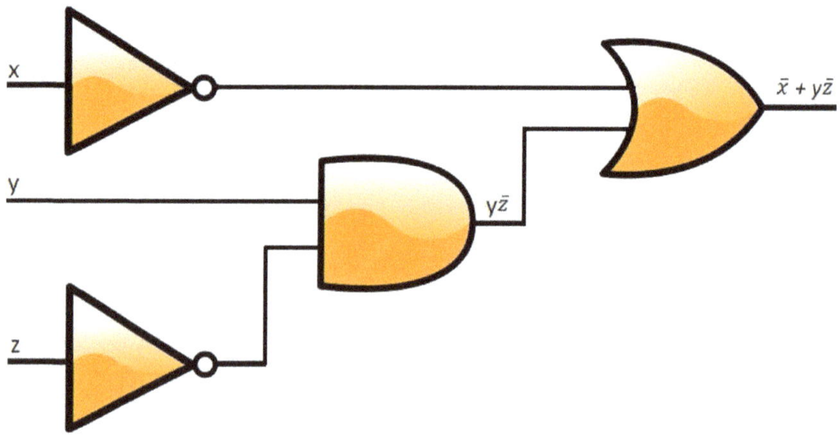

FIGURE 6.9 A Logic Diagram for $F(x, y, z) = \bar{x} + y\bar{z}$

According to the relationship between Boolean algebra and logic diagrams, digital circuits can be simplified by simplifying the corresponding Boolean expressions. As we know, gates are applied to the implementation of digital circuits. However, compared with Boolean expressions, gates and logic diagrams are not always convenient to represent digital circuits in the design process. Boolean expressions are easy to manipulate and simplify.

As shown in Examples 6.10 and 6.11, according to the given truth table, the corresponding Boolean expression is first obtained in sum-of-products form, then the logic diagram is drawn, and finally the originally drawn logic diagram is simplified, and the simplified logic diagram is drawn.

Example 6.10

TABLE 6.17 The Truth Table of the Function

X	Y	Z	F	PRODUCT
0	0	0	1	$\bar{x}\bar{y}\bar{z}$
0	0	1	0	
0	1	0	1	$\bar{x}y\bar{z}$
0	1	1	0	
1	0	0	0	
1	0	1	1	$x\bar{y}z$
1	1	0	1	$xy\bar{z}$
1	1	1	1	xyz

Given the truth table shown in Table 6.17, the Boolean expression $F(x, y, z) = \bar{x}\bar{y}\bar{z} + \bar{x}y\bar{z} + x\bar{y}z + xy\bar{z} + xyz$ is obtained, and its logic diagram is given in Figure 6.10.

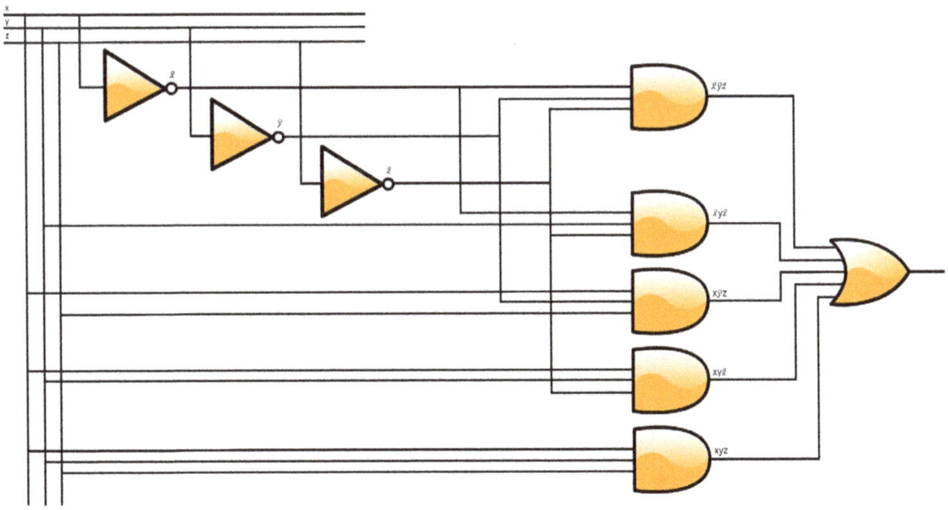

FIGURE 6.10 A Logic Diagram for $F(x, y, z) = \bar{x}\bar{y}\bar{z} + \bar{x}y\bar{z} + x\bar{y}z + xy\bar{z} + xyz$

The simplification process of F(x, y, z) is given as follows, and the simplified logic diagram is shown in Figure 6.11.

F(x, y, z) $= \bar{x}\bar{y}\bar{z} + \bar{x}y\bar{z} + x\bar{y}z + xy\bar{z} + \mathbf{xyz}$

$= \bar{x}\bar{y}\bar{z} + \bar{x}y\bar{z} + x\bar{y}z + xy\bar{z} + \mathbf{xyz} + xyz$ (according to the idempotent law; see the proof below)

$= \bar{x}\bar{y}\bar{z} + \bar{x}y\bar{z} + xy\bar{z} + xyz + xyz + x\bar{y}z$ (rearrange the items)

$= \bar{x}\bar{z}\,(\bar{y} + y) + xy(\bar{z} + z) + xz(y + \bar{y})$

$= \bar{x}\bar{z} + xy + xz$

Proof of "Adding/removing any duplicated item won't change the result."

F(x, y, z) $= xyz + xyz$

$= x(yz + yz)$

$= x(y(z + z))$

$= xyz$ (idempotent law)

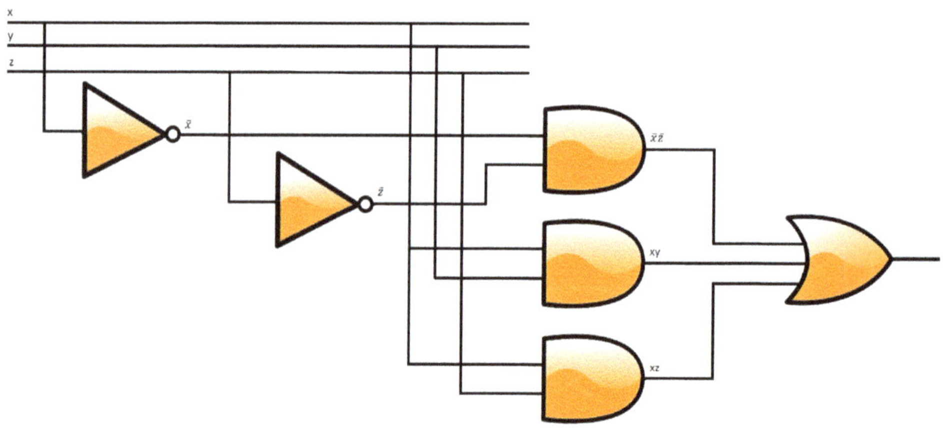

FIGURE 6.11 A Logic Diagram for the Simplified F(x, y, z) $= \bar{x}\bar{z} + xy + xz$

According to Figures 6.10 and 6.11, the Boolean expression F(x, y, z) $= \bar{x}\bar{y}\bar{z} + \bar{x}y\bar{z} + x\bar{y}z + xy\bar{z} + xyz$ is simplified to F(x, y, z) $= \bar{x}\bar{z} + xy + xz$. Nine gates are reduced to six gates. As shown in Figure 6.10, five gates involve three inputs from other gates and/or parameters, and one gate involves the inputs from the other five gates. As shown in Figure 6.11, three gates involve two inputs from other gates and/or parameters, and one gate involves the inputs from the other three gates after simplification.

Example 6.11

TABLE 6.18 The Truth Table of the Function

X	Y	Z	F	PRODUCT
0	0	0	1	$\overline{x}\overline{y}\overline{z}$
0	0	1	1	$\overline{x}\overline{y}z$
0	1	0	1	$\overline{x}y\overline{z}$
0	1	1	0	
1	0	0	1	$x\overline{y}\overline{z}$
1	0	1	1	$x\overline{y}z$
1	1	0	0	
1	1	1	0	

Given the truth table shown in Table 6.18, the Boolean expression $F(x, y, z) = \overline{x}\overline{y}\overline{z} + \overline{x}\overline{y}z + \overline{x}y\overline{z} + x\overline{y}\overline{z} + x\overline{y}z$ is obtained, and its logic diagram is given in Figure 6.12.

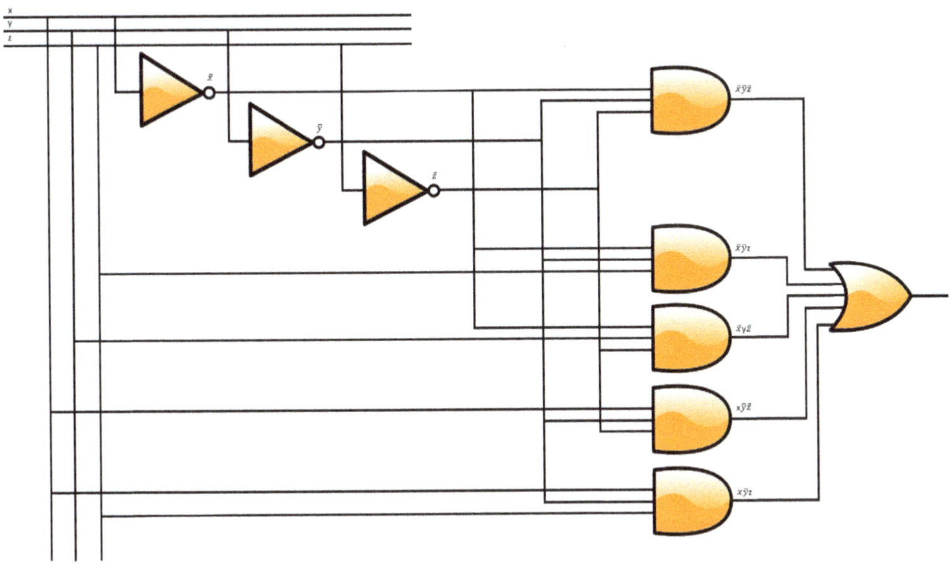

FIGURE 6.12 A Logic Diagram for $F(x, y, z) = \overline{x}\overline{y}\overline{z} + \overline{x}\overline{y}z + \overline{x}y\overline{z} + x\overline{y}\overline{z} + x\overline{y}z$

The simplification process of F(x, y, z) is given as follows, and the simplified logic diagram is shown in Figure 6.13.

$$
\begin{aligned}
F(x, y, z) &= \bar{\mathbf{x}}\bar{\mathbf{y}}\bar{\mathbf{z}} + \bar{x}\bar{y}z + \bar{x}y\bar{z} + x\bar{y}\bar{z} + x\bar{y}z \\
&= \bar{\mathbf{x}}\bar{\mathbf{y}}\bar{\mathbf{z}} + \bar{x}\bar{y}z + \bar{x}y\bar{z} + x\bar{y}\bar{z} + x\bar{y}z + \bar{\mathbf{x}}\bar{\mathbf{y}}\bar{\mathbf{z}} \\
&= \bar{x}\bar{y}\bar{z} + \bar{x}\bar{y}z + x\bar{y}\bar{z} + x\bar{y}z + \bar{\mathbf{x}}\bar{\mathbf{y}}\bar{\mathbf{z}} + \bar{x}y\bar{z} \text{ (rearrange the items)} \\
&= \bar{x}\bar{y}(\bar{z} + z) + x\bar{y}(\bar{z} + z) + \bar{x}\bar{z}(\bar{y} + y) \\
&= \bar{x}\bar{y} + x\bar{y} + \bar{x}\bar{z} \\
&= \bar{y}(\bar{x} + x) + \bar{x}\bar{z} \\
&= \bar{y} + \bar{x}\bar{z}
\end{aligned}
$$

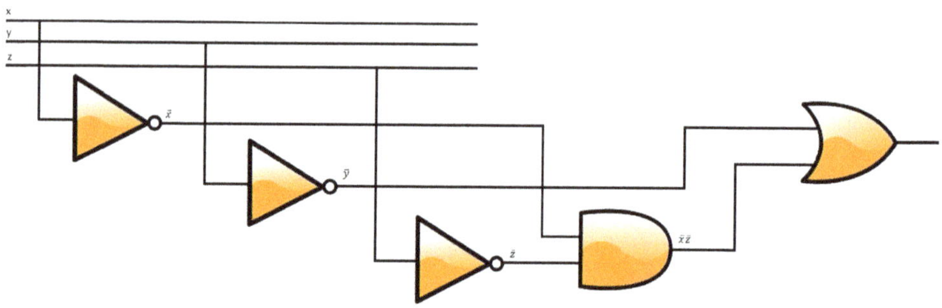

FIGURE 6.13 A Logic Diagram for the Simplified $F(x, y, z) = \bar{y} + \bar{x}\bar{z}$

According to Figures 6.12 and 6.13, the Boolean expression $F(x, y, z) = \bar{x}\bar{y}\bar{z} + \bar{x}\bar{y}z + \bar{x}y\bar{z} + x\bar{y}\bar{z} + x\bar{y}z$ is simplified to $F(x, y, z) = \bar{y} + \bar{x}\bar{z}$. Nine gates are reduced to five gates. As shown in Figure 6.12, five gates involve three inputs from other gates and/or parameters, and one gate involves the inputs from the other five gates. As shown in Figure 6.13, two gates involve the inputs from the other two gates after simplification.

The complexity of the Boolean expression directly determines the complexity of the corresponding digital circuit. A complex Boolean expression generates a complex digital circuit. How do we simplify digital circuits? Sometimes simplifying Boolean expressions is quite difficult and time-consuming. So, *Karnaugh maps* (or *K-maps*) are introduced rather than directly using Boolean algebra to simplify digital circuits. The related details are discussed later.

Chapter Summary

Three Boolean operations (AND, OR, and NOT), Boolean expressions, gates, logic diagrams, and basic simplification of Boolean expressions are discussed in this chapter. The formal definitions of Boolean expressions and truth tables are given. Multiple examples are used to illustrate the conversion among Boolean expressions, truth tables, and logic diagrams. According to the basic identities of Boolean algebra, the necessity of Boolean expression simplification is explained, and the corresponding Boolean expression simplification process is illustrated by formal analysis and logic diagrams.

Students should be able to achieve the conversion between truth tables and Boolean expressions and draw the corresponding logic diagrams. Students can complete the basic simplification of Boolean expressions. The advanced simplification of Boolean expressions will be further discussed in the next chapter.

Chapter 6 Exercises

1. Fill in the truth table for the following Boolean expression:

$$f(x, y, z) = (\overline{x} + y)(y + z)$$

X	Y	Z	$\overline{X}$	$(\overline{X} + Y)$	$(Y + Z)$	$(\overline{X} + Y)(Y + Z)$

2. Draw a digital logic diagram for the function in problem 1.

3. Construct a truth table to prove the OR form of the distributive identity for Boolean algebra:

$$x(y + z) = xy + xz$$

Include columns computing $(y + z)$, xy, xz, $x(y + z)$, and $x(y + z)$.

4. Indicate the Boolean identity used in each of the following four steps reducing the expression:

$$a(b + a + \overline{b})$$
$$= a(a + b + \overline{b})$$
$$= a(a + 1)$$
$$= a(a)$$
$$= a$$

5. Using the basic identities of Boolean algebra to prove that:

   a. $xy + x\overline{y} = x$
   b. $x + \overline{x}y = x + y$
   c. $x + \overline{x}z + yz = xy + \overline{x}z$

6. Fill in a four-variable truth table equivalent to the following digital logic diagram.

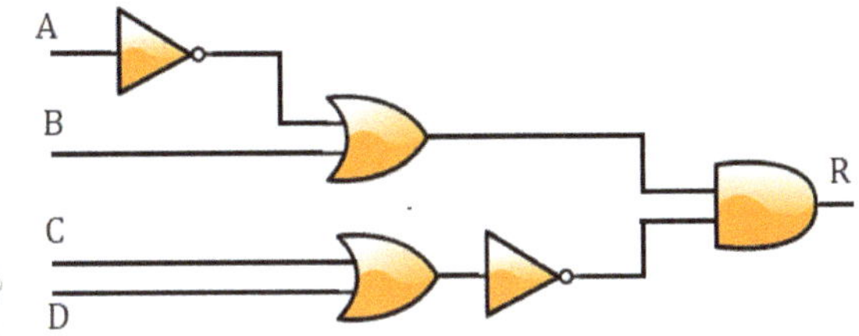

Image 6.1

7. Identify the logical expressions of the following digital logic diagrams.

   a.

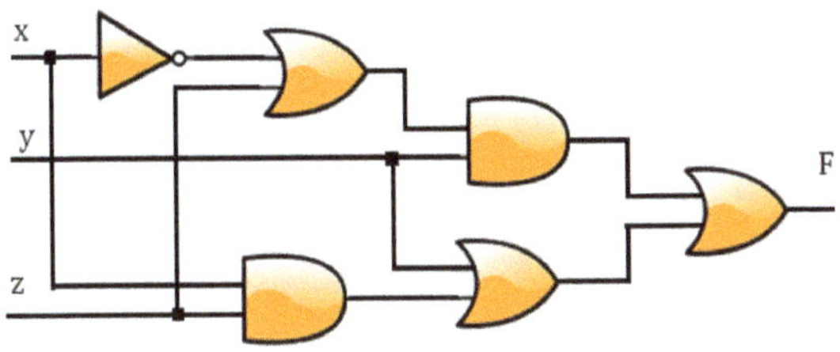

Image 6.2a

   b.

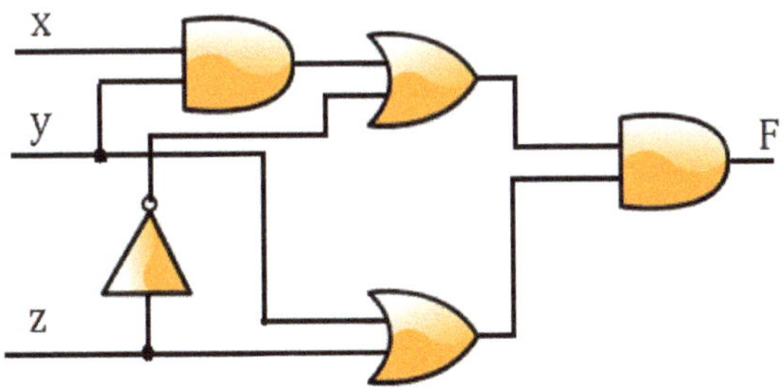

Image 6.2b

CHAPTER

7

Karnaugh Maps

Introduction

In Chapter 6, we have discussed the relationship of Boolean expressions and digital logic circuits. Each digital circuit representation is determined by its Boolean expression. Most "real-life" circuits are quite large and complex if the expressions are not simplified. These un-simplified circuits have unnecessary digital elements such as transistors and wires, which would consume more power and perform slower.

Simplification of a Boolean expression leads to simpler, faster, and energy-saving digital circuits. By applying laws of Boolean algebra, a Boolean expression can be reduced and thus minimize the complexity of combinational and sequential circuits. However, simplifying Boolean functions using laws and rules of Boolean identities is time-consuming and error-prone. A Karnaugh map (K-map) is an easy, systematic, and programmable method for reducing Boolean expressions. In this chapter, K-maps are introduced to simplify Boolean expressions.

Objectives

This chapter will help students:

- **understand** the motivation of using a K-map for expression simplification.
- **convert** a truth table into a K-map.

- **apply** rules to group cells in a K-map.
- **produce** the minimum product terms from the groups.
- **generate** the simplified Boolean expression from a given truth table.
- **identify** the un-simplified Boolean expression.

Key Terms

Gray code sequence: sequence of binary words of a given length where each successive word in the sequence varies in value in only 1-bit position.

contradictory variable: it is a variable whose literals occurring both complemented and uncomplemented forms within the minterm list of a K-map group. Contradictory variables must be eliminated in the expression.

minterm: a product (AND) term consisting of all the variable literals, either complemented or uncomplemented.

maxterm: a sum-up (OR) term consisting of all the variable literals, either complemented or uncomplemented.

minimum product term: after the contradictory variables are eliminated from the minterm list, the survived variables are in a multiplied (AND) form.

Motivation for Adopting the Karnaugh Map

As we know, according to Moore's law, the number of transistors on a microchip doubles every 2 years, increasing the speed and capability of computers. However, this development trend is not going to be boundless. There are limits to Moore's law. Denser chips and faster signal processing both generate increased heat. Current hardware technology is right at the edge of this boundary with a high density of gates (transistors) generating high enough heat to melt the chips and destroy the computer.

Developing a powerful chip with low cost is always a challenging task for the hardware designers. Boolean expression simplification can be used to decrease the density of transistors on a chip. Digital circuits are designed according to the Boolean expression. Without optimization, Boolean expressions can be complex and redundant, which leads to overly complex digital circuits. Removing the unnecessary components in the circuit is always a crucial step in the process of digital circuit design. The simpler the Boolean expression, the simpler the resulting circuit will be (cheaper, consumes less power, and faster). With this in mind, we always want to reduce the Boolean functions to their simplest form.

In Chapter 6, Boolean identity laws were used for Boolean expression simplification. However, using Boolean algebra laws, rules, and theorems to simplify logic expression is tedious, difficult, and error-prone, especially for functions with dozens or even hundreds of variables. It is almost impossible to apply this approach for a very large integrated circuit design (with hundreds and millions of Boolean operations).

Maurice Karnaugh was a telecommunication engineer at Bell Labs. In 1953, while exploring the new field of digital logic and its application to the design of telephone circuits, he invented a graphical way of visualizing and then simplifying Boolean expressions. This graphical representation, now known as a Karnaugh map, or K-map, is named in his honor.

A K-map provides a systematic method for simplifying Boolean expressions. If properly used, it will produce the minimum Boolean expression.

Karnaugh Map Structure

A K-map is derived from a regular truth table. It provides a different representation of the outputs of the logical expression. The map is a 2-D table (matrix) consisting of rows and columns that represent the output values of a Boolean function. The output values placed in each cell are derived from the minterms of a Boolean function. A *minterm* is a product (AND) term consisting of all the variable literals, either complemented or not complemented.

To make this map meaningful to beginners, in this book, a top row and a leftmost column are added into the map as the horizontal and vertical axes of the table structure, respectively. The values on both axes are from the binary values of the input variables. The structure of the K-map is like a spreadsheet (MS Excel). Each cell in this "spreadsheet" has its row and column coordinates, which are the binary values of the input variable. The cell value is the output from the truth table determined by the corresponding input variable values.

Let us look at the following example:

Example 7.1: Derive a K-map from a truth table (left) with two variables *x* and *y*; *F* is the output.

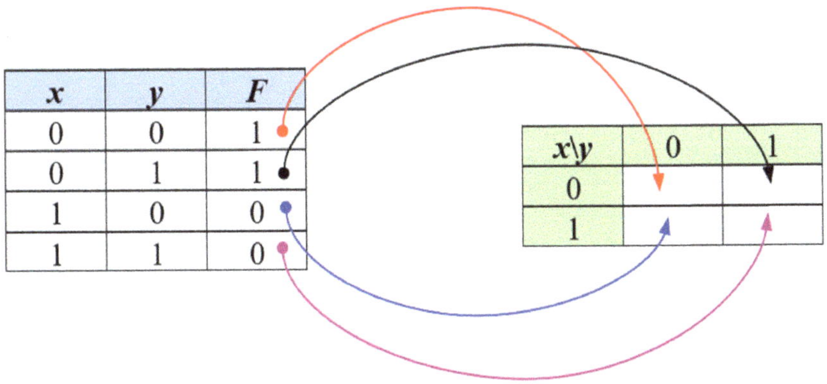

Image 7.1

Its corresponding K-map structure is on the right side, where "x\y" in the top left cell indicates what the values of the first row and the first column (the leftmost) would be. For this example, *x* values are used as the number on the vertical axis, while *y* values are horizontal coordinates. The arrows in this example indicate the source of the K-map cell values. The red arrow shows that the output ***F*** value 1 from $x = 0$, $y = 0$ in the truth table would be in the K-map's cell whose coordinate is $x = 0$ and $y = 0$, respectively. Similarly, the other arrows indicate the positions of the other three output ***F*** values. The final K-map is:

X\Y	0	1
0	1	1
1	0	0

The above example is the simplest K-map with only two variables. The coordinate values of vertical and horizontal axes are just 0 and 1, respectively. Let us see another example which has more than two variables.

Example 7.2: Derive a K-map from a truth table (left) with three variables; F is the output.

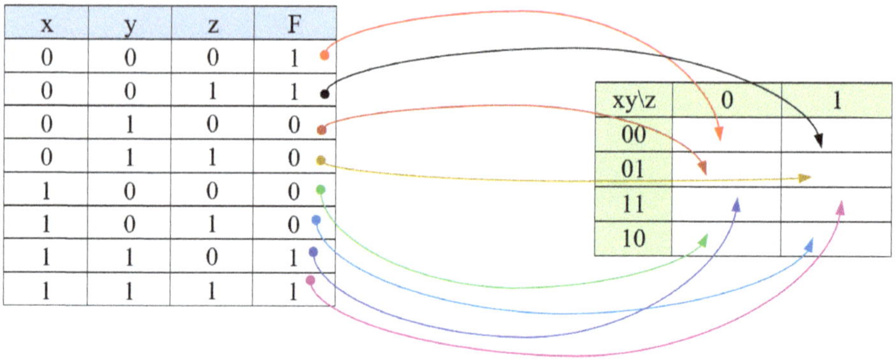

x	y	z	F
0	0	0	1
0	0	1	1
0	1	0	0
0	1	1	0
1	0	0	0
1	0	1	0
1	1	0	1
1	1	1	1

K-map:

xy\z	0	1
00	1	1
01	0	0
11	1	1
10	0	0

image 7.2

This K-map for three variables is still a 2-D table, but either one of the axes would have to contain the values from more than one variable. Here, *x* and *y* values are combined to be the vertical coordinates, while the *z* variable represents the horizontal values. The final K-map cells are in a table with two columns and four rows.

For the same truth table in **Example 7.2**, the K-map could also be organized in a table of four columns and two rows, where instead of **x** and **y**, **y** and **z** are combined for the horizontal coordinate. The following is the result for this scenario.

Example 7.3: From a truth table to a K-map: For a three-variable truth table, F is the output.

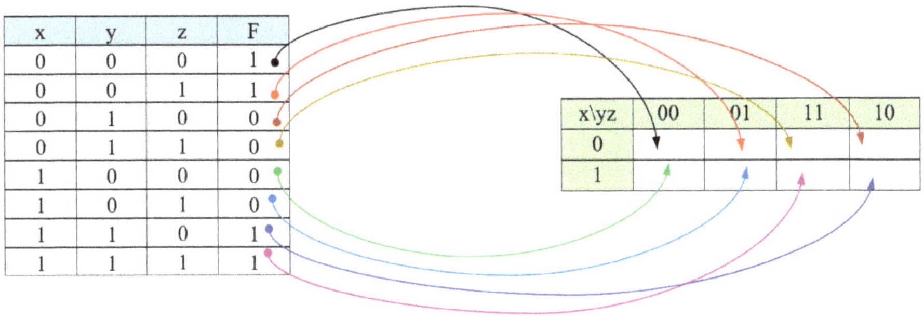

x	y	z	F
0	0	0	1
0	0	1	1
0	1	0	0
0	1	1	0
1	0	0	0
1	0	1	0
1	1	0	1
1	1	1	1

x\yz	00	01	11	10
0				
1				

K-map:

x\yz	00	01	11	10
0	1	1	0	0
1	0	0	1	1

Image 7.3

Note: In **Example 7.3** and **Example 7.4**, the sequences of numbers across the top row or the leftmost column of the K-maps are not in binary order, like 00, 01, 10, 11. Instead, they are 00, 01, 11, and 10 for xy in Example 7.2, and for yz in Example 7.3. The binary data in such order is called a **Gray code sequence**. A Gray code sequence only changes one binary bit as we go from one value to the next in the sequence.

TABLE 7.1 Gray and Binary Code Sequences for 2-bit Binaries

Gray code sequence

BIT 1	BIT 2
0	0
0	1
1	1
1	0

Binary code sequence

BIT 1	BIT 2
0	0
0	1
1	0
1	1

If we check the sequence from top to bottom in the left-side table in Table 7.1, from the first row to the second, bit 2 is changed from 0 to 1. From the second to the third row, bit 1 is changed from 0 to 1. From the third to the fourth row, only bit 2 again is changed from 1 to 0. This is the so-called Gray code sequence. On the right-side table in Table 7.1, a regular binary code counting sequence is shown, where both bit values are changed, moving from the second to the third rows. This does not satisfy the single bit change of a Gray code sequence.

By enforcing the Gray code sequence in a K-map, the adjacent cells will only vary by 1 bit, or Boolean variable. This is to organize the outputs of a logic function in a K-map so that we may view the commonality between variables.

Again, the column and row headings of a K-map must be in Gray code order. Thus, adjacent cells in K-map vary by only 1 bit because a Gray code sequence varies by only 1 bit. Without following the Gray code sequence, cells sharing common Boolean variables would no longer be adjacent or show visual patterns, and the map will not work appropriately for the purpose of expression simplification.

Example 7.4: From a truth table to a K-map: For a four-variable truth table, F is the output.

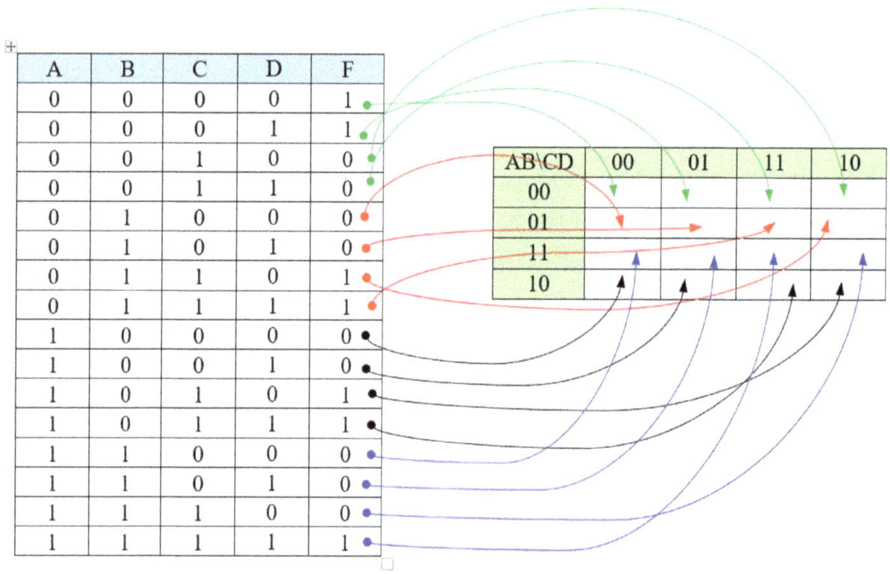

A	B	C	D	F
0	0	0	0	1
0	0	0	1	1
0	0	1	0	0
0	0	1	1	0
0	1	0	0	0
0	1	0	1	0
0	1	1	0	1
0	1	1	1	1
1	0	0	0	0
1	0	0	1	0
1	0	1	0	1
1	0	1	1	1
1	1	0	0	0
1	1	0	1	0
1	1	1	0	0
1	1	1	1	1

K-map:

AB\CD	00	01	11	10
00	1	1	0	0
01	0	0	1	1
11	0	0	1	0
10	0	0	1	1

Image 7.4

Note: Both AB and CD variables follow the Gray code sequence order.

In Example 7.4, there are four variables: A, B, C, and D in the truth table. It is easy to group AB and CD into vertical and horizontal axes, respectively, with the values in Gray code sequence (00, 01, 11, 10). The final K-map is a 4×4 table (matrix) with 16 cells.

Simplification with Karnaugh Maps

Once a K-map has been constructed, the first step in simplifying the expression is to spot groups of adjacent 1s in the K-map cells. There are several grouping rules to follow. If the rules are violated, the grouped results will not lead to the simplest Boolean expression.

Rule 1: Only 1s are grouped; 0s are ignored.

Example 7.5: K-map grouping.

K-MAP:

x\yz	00	01	11	10
0	1	1	0	0
1	0	0	1	1

For the K-map in Example 7.5, two groups of 1s are identified, while 0s are not counted.

Rule 2: A group of 1s must be adjacent horizontally or vertically.

In Example 7.5, the diagonal neighbors of 1s in the shaded cells cannot be grouped.

Rule 3: The member size of a group must be powers of 2; e.g., 2^0, 2^1, 2^2, 2^3, 2^4, etc.

In Example 7.5, each group has two members; the size is 2^1. A group with only a cell is also possible because $2^0 = 1$.

Rule 4: Group size: the bigger the group, the better.

Example 7.6: K-map grouping.

K-MAP:

xy\z	0	1
00	1	0
01	1	0
11	1	1
10	1	1

In Example 7.6, the group within the blue dashed box with three members is **not** valid because 3 is not a number with power of 2. The member size of the black dashed box is 2, which is power of 2. However, by checking their neighbors, a bigger group with four members on the same column can be formed. Therefore, the bigger group (in red box) should be chosen, and the black group violates Rule 4. If the group within the dashed black box is mistakenly chosen, the final Boolean expression will not be the simplest.

Rule 5: The shape of a group must be a regular rectangle/square.

Example 7.7: Incorrect group with irregular shape in K-map grouping.

K-MAP:

xy\z	0	1
00	1	0
01	1	0
11	1	1
10	0	0

In Example 7.7, the group surrounded by the red dashed line violates Rule 5, as the group shape is not a regular rectangle.

Rule 6: A single "1" may occur in more than one group if it can make a bigger group.

Example 7.8: K-map grouping.

K-MAP:

xy\z	0	1
00	1	0
01	1	0
11	1	1
10	1	0

In Example 7.8, the blue group is overlapped with the red group. It is valid as the shaded cell belongs to both groups according to Rule 6. Without including this shaded cell, the blue group has a single member, which seems valid according to the previous rules. However, by including the overlapped cell into the blue group, the bigger group will lead to a simpler final expression.

Rule 7: Every 1 must occur in a group.

Example 7.9: K-map grouping.

K-MAP:

xy\z	0	1
00	1	0
01	1	0
11	0	0
10	0	1

In Example 7.9, there is no adjacent neighbor for the element in the blue box. In this case, a single member group has to be formed for this isolated cell. The group size is $2^0 = 1$, which still follows Rule 3.

Rule 8: Group members can be the neighbors that are wrapped around the rows and columns.

Example 7.10: K-map grouping with wraparound neighbors.

K-MAP:

xy\z	0	1
00	1	1
01	0	0
11	0	0
10	1	1

The cells at both ends of the columns or rows are considered as adjacent neighbors. They can be grouped if other grouping rules are applicable. In Example 7.10, four corner cells are all 1s and can be grouped together.

Let us do grouping in the K-map in Example 7.11.

Example 7.11: K-map grouping.

K-MAP:

AB\CD	00	01	11	10
00	1	0	0	1
01	1	0	1	1
11	1	1	0	0
10	1	1	0	0

There are three groups that can be formed according to the rules:

- Green group: a square shape, adjacent neighbors with 1's; it has four (2^2) members.
- Blue group: two adjacent members in a row.
- **Black group**: this is a wraparound group. Four members are wraparound in the horizontal direction. Meanwhile, this group is partially overlapped with both the red and blue groups.

Rule 9: The number of groups should be as minimal as possible.

Note: In Example 7.11, the dashed red box is NOT a valid group. Its top two cells are overlapped with the black group, and the bottom two cells are intersected with the green group. All cells in the red box have been covered by other groups. There is no contribution from this group to cover any ungrouped cells. Therefore, three groups in Example 7.11 are enough, and this dashed red group is redundant.

Minimum Expression from a Karnaugh Map

Based on the groups identified on a K-map, the minimum logical expression of sum-of-product can be formed. The sum-of-product form is introduced in Chapter 6. Basically, it is the summation of several minimum product terms, which is related to the minterms introduced in the Chapter 6. Each group identified on the K-map corresponds to one minimum product term.

For a Boolean function, there are two types of variable combinations: maxterm and **minterm**. A maxterm is the form that variable name-related literals are summed up (OR), while a minterm is the form that variable literals are multiplied (AND), sometimes also called a product term. The literal format depends on the variable Boolean values:

VARIABLE VALUE	LITERAL FORM
True (1)	variable name's uncomplemented literal
False (0)	variable name's complemented literal

Table 7.2 lists all maxterms and minterms for two Boolean variables.

TABLE 7.2 Maxterms and Minterms for Two Boolean Variables

X	Y	MAXTERM	MINTERM
0	0	$x' + y'$	$x'y'$
0	1	$x' + y$	$x'y$
1	0	$x + y'$	xy'
1	1	$x + y$	xy

Similarly, maxterms and minterms for three Boolean variables are in Table 7.3.

TABLE 7.3 Maxterms and Minterms for Three Boolean Variables

X	Y	Z	MAXTERM	MINTERM
0	0	0	$x' + y' + z'$	$x'y'z'$
0	0	1	$x' + y' + z$	$x'y'z$
0	1	0	$x' + y + z'$	$x'yz'$
0	1	1	$x' + y + z$	$x'yz$
1	0	0	$x + y' + z'$	$xy'z'$
1	0	1	$x + y' + z$	$xy'z$
1	1	0	$x + y + z'$	xyz'
1	1	1	$x + y + z$	xyz

From a group identified on a K-map, there are several members. For each group member (cell), we can have a minterm according to its variable values. Figure 7.1 shows the K-map of Example 7.8, where the group in the red box has four members.

xy\z	0	1
00	1	0
01	1	0
11	1	1
10	1	0

Coordinate	minterm
00 0	x'y'z'
01 0	x'yz'
11 0	xyz'
10 0	xy'z'

FIGURE 7.1 Minterms for a K-map Group

These four cells' coordinates deriving from the variables *x*, *y*, and *z* from the top to bottom are 000, 010, 110, and 100, respectively, and the corresponding minterm list is *x'y'z'*, *x'yz'*, *xyz'*, and *xy'z'* (shown in the right table in Figure 7.1). Within this minterm list, if a certain variable's literal occurs in both complemented and uncomplemented forms, it is called a **contradictory variable**. In this example, both *x* and *y* are the *contradictory variables* because *x'* (complemented literal of variable *x*) appears in the first two rows; meanwhile, *x* (uncomplemented literal) also appears in the last two rows. Similarly, *y'* appears in the first and the last rows, and *y* appears in the middle two rows. *Contradictory variables* must be eliminated in the expression for this group, and the survived literals are the **minimum product term**. For the group in Figure 7.1, the minimum product term is *z'*. This process is repeated for every group. When all the *minimum product terms* are derived from the K-map, their summed-up form is the minimum sum-of-product expression for this K-map.

Example 7.12: Let us see the grouped results from Example 7.11.

There are three groups.

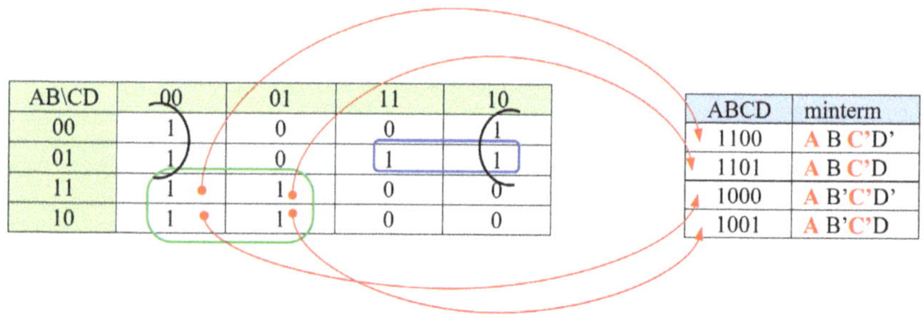

Image 7.5

The green group has four members whose ABCD variable values and corresponding minterms are on the right side. Here, variables B and D have both complemented and uncomplemented forms; e.g., AB'C'D', ABC'D. According to the rule, B and D are *contradictory variables* and need to be eliminated. Variables A and C keep the consistent form across over four cells, where A are all uncomplemented and C are all complemented. Thus, the minimum product term for the green group is **AC'**. If we look pictorially at the K-map (circled group in green), we can see this idea by asking the question: which variables do not change in the group? We can see that B and D change between 0 and 1 in the group, but A and C remain constant with values of 1 and 0, respectively (indicating A and C').

The contradictory variable of the blue group is just D, whose value is changed from 0 to 1:

ABCD	MINTERM
0110	A'B CD'
0111	A'B CD

Therefore, the minimum product term of the blue group is **A'BC**.

The wraparound group within the black line has four members, and their variable values and corresponding **minterms** are:

ABCD	MINTERM
0000	A'B' C'D'
0010	A'B' CD'
0100	A' B C'D'
0110	A' B CD'

Both variables A and D are found with consistent complemented form, while others are contradictory variables. The minimum product term for this group is A'D'.

Finally, the simplified expression from this K-map is the sum of these minimum product terms:

F(A,B,C,D) = AC' + A'BC + **A'D'**

Let us see another case in Example 7.13.

Example 7.13

Using a K-map, provide the simplified Boolean expression according to the following truth table:

ABCD	F(A,B,C,D)	ABCD	F(A,B,C,D)
0000	0	1000	0
0001	1	1001	1
0010	1	1010	0
0011	0	1011	0
0100	0	1100	0
0101	1	1101	1
0110	1	1110	1
0111	1	1111	1

Step 1: K-map construction and grouping

AB\CD	00	01	11	10
00	0	1	0	1
01	0	1	1	1
11	0	1	1	1
10	0	1	0	0

Step 2: Sum of minimum product terms is the simplified Boolean expression for the truth table:

C'D + BC + A'CD'

Chapter Summary

Previously, Boolean identity laws are used for Boolean expression simplification. However, using Boolean algebra laws, rules, and theorems to simplify a logic expression is tedious, difficult, and error-prone. In this chapter, a systematic Boolean expression simplification method, a Karnaugh map (K-map), is introduced. A K-map is derived from a regular truth table. It provides a different structure, a 2-D table (matrix) consisting of rows and columns that represent the output values of a Boolean function.

Gray code sequence is a key to organizing a K-map that leads to the simplified Boolean expression. Grouping on a K-map is done by applying nine rules. Each cell in an identified group has a minterm. By eliminating the contradictory variables, a minimal product term is available for each group. The final simplified expression is the summed minimal product terms for all groups.

The following is the complete simplification process:

1. Construct the K-map and place 1s and 0s in the squares according to the truth table.
2. Spot groups of adjacent 1s in the K-map cells according to a set of rules:

- **Rule 1:** Only 1s are grouped; 0s are ignored.
- **Rule 2:** A group of 1s must be adjacent horizontally or vertically.
- **Rule 3:** The member size of a group must be powers of 2; e.g., 2^0, 2^1, 2^2, 2^3, 2^4, etc.
- **Rule 4:** Group size: the bigger the group, the better.

- **Rule 5:** The shape of a group must be a regular rectangle/square.
- **Rule 6**: A single "1" may occur in more than one group if it can make a bigger group.
- **Rule 7:** Every 1 must occur in a group.
- **Rule 8**: Group members can be the neighbors that are wrapped around the rows and columns.
- **Rule 9**: The number of groups should be as minimal as possible.

3. Form a minterm for each K-map group.
4. Form the OR sum of all the minterms generated by individual groups.

Chapter 7 Exercises

1. Convert the following truth tables into K-maps and provide the simplified expressions.

a.

x	y	z	F
0	0	0	1
0	0	1	0
0	1	0	1
0	1	1	0
1	0	0	0
1	0	1	1
1	1	0	1
1	1	1	1

b.

x	y	z	F
0	0	0	1
0	0	1	1
0	1	0	1
0	1	1	0
1	0	0	1
1	0	1	1
1	1	0	0
1	1	1	0

c.

A	B	C	D	F
0	0	0	0	0
0	0	0	1	1
0	0	1	0	1
0	0	1	1	0
0	1	0	0	0
0	1	0	1	1
0	1	1	0	1
0	1	1	1	1
1	0	0	0	0
1	0	0	1	1
1	0	1	0	0
1	0	1	1	0
1	1	0	0	0
1	1	0	1	1
1	1	1	0	1
1	1	1	1	1

2. Provide simplified logical expressions for the following K-maps:

a.

x|yz	00	01	11	10
0	0	1	1	0
1	1	0	0	1

b.

x|yz	00	01	11	10
0	0	1	1	1
1	1	0	0	0

x|yz	00	01	11	10
0	1	1	1	0
1	1	1	1	1

c.

x|yz	00	01	11	10
0	1	0	0	1
1	1	0	1	1

d.

wx|yz	00	01	11	10
00	0	1	0	1
01	0	1	1	1
11	0	1	1	0
10	0	1	0	0

CHAPTER

8

Digital Circuit Design

Introduction

This chapter introduces integrated circuits (ICs), discusses the basic combinational circuit design, including the adder, decoder, multiplexer, parity generator, and parity checker, and covers basic sequential circuit design, including the clock and flip-flops. Logic NAND gates are used as the basic "building blocks" to construct other logic gate functions and even implement complex ICs. The implementation of a half-adder and a full-adder is discussed in detail. The implementation of a 16-bit adder using full-adders is also covered. The detailed analysis of decoders, multiplexers, parity generators, and parity checkers is given and illustrated by using Boolean expressions, truth tables, and logic diagrams. The basic concepts of the clock and flip-flops are given and explained using different examples.

Objectives

This chapter will help students:

- **understand** integrated circuits, combinatorial circuits (including adder, decoder, multiplexer, parity generator, and parity checker), and sequential circuits (including clock and flip-flop).
- **master** the design and implementation of adder, decoder, and multiplexer, clock, and flip-flop.
- **demonstrate** the basic design of combinatorial circuits and sequential circuits.

Key Terms

adder: a digital circuit that performs addition of numbers.

decoder: a combinational circuit that converts *n* lines of input into 2*n* lines of output.

multiplexer: a combinational logic circuit that selects between several analog or digital input signals and forwards the selected input to a single output line.

parity generator: a combinational logic circuit that generates the parity bit in the transmitter.

parity check: the process that ensures accurate data transmission between nodes during communication.

clock: a circuit that provides a sequence of pulses to trigger each internal operation.

flip-flop: an electronic circuit with two stable states that can be used to store binary data.

Introduction to the Integrated Circuit

Chapter 7 discussed various gates. Gates can be connected to build computers. The collections of gates are used to build blocks, then the blocks are used to build the entire computer system. Three basic AND, OR, and NOT operations are applied to the construction of blocks. Various digital components are connected by wires to compose computers. Gates are combined in different ways to implement various functions. How can different gates work together to implement complex functions? This is done through the design of integrated circuits (ICs). As a miniature and low-cost electronic circuit, an IC consists of a set of active and passive components integrated on a single crystal of silicon. The active components include transistors and diodes. The passive components include resistors and capacitors.

ICs have seven main advantages:

1. miniaturization and hence increased equipment density
2. cost reduction due to batch processing
3. increased system reliability due to the elimination of soldered joints
4. improved functional performance
5. matched devices

6. increased operating speeds
7. reduction in power consumption

Depending upon the number of active devices per chip, there are different levels of integration, as shown in Table 8.1. A simple SSI integrated circuit is shown in Figure 8.1.

TABLE 8.1 The Level of Integration in an IC

	LEVEL OF INTEGRATION	NUMBER OF ACTIVE DEVICES PER CHIP
1	Small-scale integration (SSI)	Less than100
2	Medium-scale integration (MSI)	100–10,000
3	Large-scale integration (LSI)	1000–100,000
4	Very large-scale integration (VLSI)	Over 100,000
5	Ultra-large-scale integration (ULSI)	Over 1 million

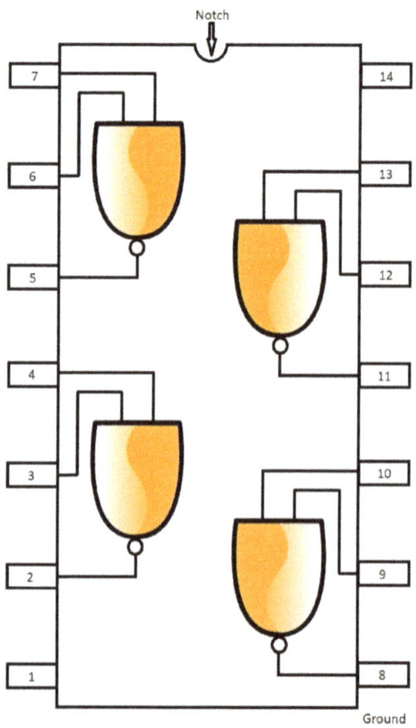

FIGURE 8.1 A Simple SSI Integrated Circuit

Combinational Circuits

The logic circuits composed by the combinations of digital logic chips can be categorized as either *combinational logic* or *sequential logic*.

Basic Concepts

Circuits containing basic Boolean operators, inputs, and outputs are built by combinational logic. The output of a combinational circuit is determined by the function of inputs and the input values. A combinational circuit is not limited to one output. If several outputs are involved, each output corresponds to a different Boolean function.

Examples of Typical Combinational Circuits

A *half-adder* is a simple combinational circuit to add two binary digits together. In the binary system, there are only three ways to add two binary numbers: 0 + 0 = 0, 0 + 1 = 1 + 0 = 1, and 1 + 1 = 10. According to the truth table for a half-adder shown in Table 8.2, the sum has two possible outputs, either 0 or 1. Only 1 + 1 involves a carry.

TABLE 8.2 The Truth Table for a Half-Adder

INPUTS		OUTPUTS	
x	*y*	SUM	CARRY
0	0	0	0
0	1	1	0
1	0	1	0
1	1	0	1

As shown in Figure 8.2, the logic diagram of a half-adder can be implemented by an XOR gate and an AND gate. An XOR gate and an AND gate are used to implement sum output and carry output, respectively.

A half-adder can only add two bits together, so it is not very useful. How do we extend the half-adder and apply it to the addition of large binary numbers in digital circuits? Let's use the addition of base-10 numbers as an example. The addition starts from the rightmost column. The rightmost columns are added to give a sum, and perhaps a non-zero carry. Then, the non-zero carry is added to the second-to-the-rightmost columns. The process is repeated until the addition of the leftmost column is completed. The addition of binary numbers is processed in the same way.

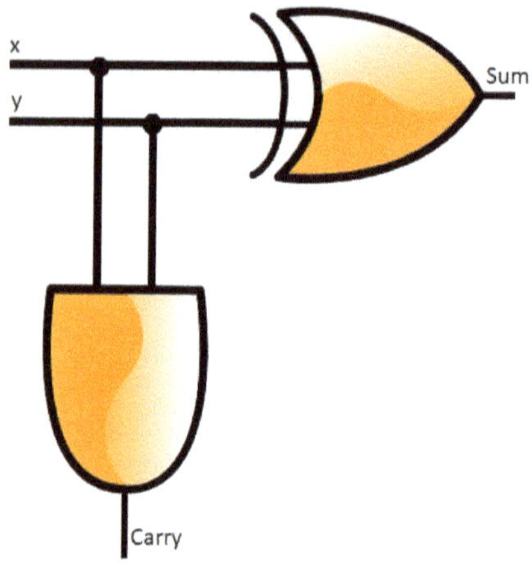

FIGURE 8.2 The Logic Diagram for a Half-Adder

Table 8.3 and Figure 8.3 show the truth table and corresponding logic diagram for a *full-adder*, respectively. A full-adder has three inputs (x, y, and carry-in) and two outputs (sum and carry-out). Let's compare the full-adder shown in Figure 8.3 with the half-adder shown in Figure 8.2. The full-adder is composed of two half-adders and an OR gate. A complex circuit can be composed by combining multiple simple circuits.

TABLE 8.3 A Truth Table for a Full-Adder

INPUTS			OUTPUTS	
x	*y*	CARRY IN	SUM	CARRY OUT
0	0	0	0	0
0	0	1	1	0
0	1	0	1	0
0	1	1	0	1
1	0	0	1	0
1	0	1	0	1
1	1	0	0	1
1	1	1	1	1

This full-adder can only add three bits. How can the addition of binary numbers with a larger number of bits be implemented in a circuit? As mentioned above, a complex circuit can be composed of multiple simple circuits. An adder capable of adding two 16-bit numbers can be built by replicating the full-adder shown in Figure 8.3, 16 times. The carry-out of one circuit is fed into the carry-in of the next

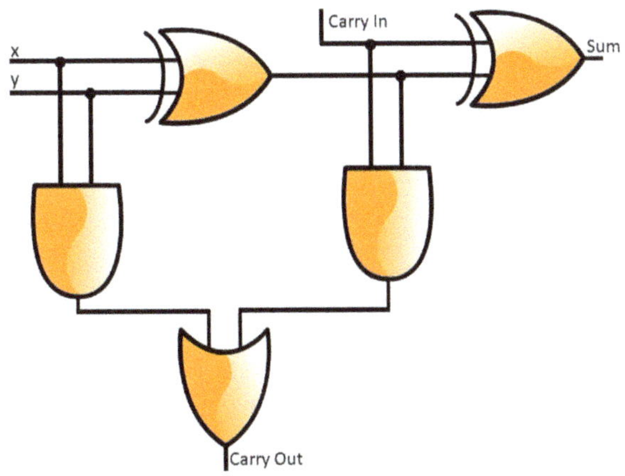

FIGURE 8.3 A Logic Diagram for a Full-Adder

circuit to its left. Figure 8.4 shows the implementation process of building a 16-bit adder, which is called a ripple-carry adder. In a ripple-carry adder, the carry-out of each full-adder is the carry-in of the succeeding full-adder through the adder stages. To simplify the drawing of a 16-bit adder, black boxes (marked as FA) without gate details are used to represent full-adders, and only the inputs and outputs of the circuits are given.

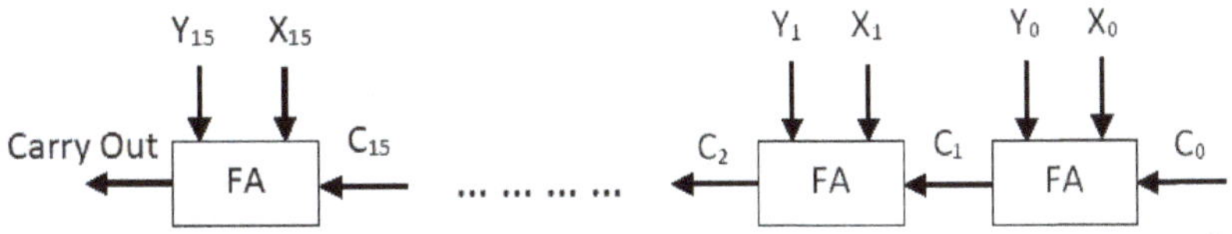

FIGURE 8.4 The Logic Diagram for a Ripple-Carry Adder

However, the processing speed of the ripple-carry adder is slow. Therefore, it is not normally used. Different modifications are made to optimize the adder performance, such as the carry-look-ahead adder, the carry-select adder, and the carry-save adder. The core idea of optimization focuses on shortening the delay required to add two binary numbers. These adders increase execution speed by not waiting for previous sums to generate carries but instead by computing carries directly from input in parallel with the sums.

Adders are important circuits for each computer. As a common operation of computers, binary information is decoded from a set of n inputs to a maximum of 2^n outputs. The inputs and their respective values are used to select one specific output line by decoders. The selection of an output line means one unique output line is set to 1, while the other output lines are set to 0. The number of inputs and the number of outputs are used to define decoders. For example, a decoder that has three inputs and eight outputs is called a 3-to-8 decoder. In computers, any memory address is specified as a binary number. In both cases of reading or writing operations related to a memory address, its actual address needs to be first determined by a decoder.

Example 8.1: A 3-to-8 decoder circuit.

As shown in Figure 8.5, a 3-to-8 decoder has three inputs (S_0, S_1, S_2) and eight outputs (D_0 to D_7). According to the truth table shown in Table 8.4, only one of eight outputs (D_0 to D_7) is selected based on three selected inputs.

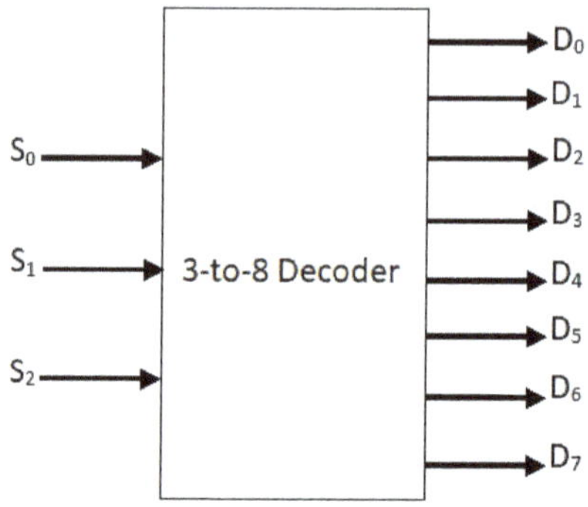

FIGURE 8.5 A 3-to-8 Decoder Block Diagram

TABLE 8.4 A 3-to-8 Decoder Truth Table

INPUTS			OUTPUTS							
S_0	S_1	S_2	D_0	D_1	D_2	D_3	D_4	D_5	D_6	D_7
0	0	0	1	0	0	0	0	0	0	0
0	0	1	0	1	0	0	0	0	0	0
0	1	0	0	0	1	0	0	0	0	0
0	1	1	0	0	0	1	0	0	0	0
1	0	0	0	0	0	0	1	0	0	0
1	0	1	0	0	0	0	0	1	0	0
1	1	0	0	0	0	0	0	0	1	0
1	1	1	0	0	0	0	0	0	0	1

According to the above truth table of a 3-to-8 decoder, the logic expression can be defined as follows:

$$D_0(S_0, S_1, S_2) = \overline{S}_0\overline{S}_1\overline{S}_2$$

$$D_1(S_0, S_1, S_2) = \overline{S}_0\overline{S}_1 S_2$$

$$D_2(S_0, S_1, S_2) = \overline{S}_0 S_1\overline{S}_2$$

$$D_3(S_0, S_1, S_2) = \overline{S}_0 S_1 S_2$$

$$D_4(S_0, S_1, S_2) = S_0\overline{S}_1\overline{S}_2$$

$$D_5(S_0, S_1, S_2) = S_0\overline{S}_1 S_2$$

$$D_6(S_0, S_1, S_2) = S_0 S_1\overline{S}_2$$

$$D_7(S_0, S_1, S_2) = S_0 S_1 S_2$$

According to the above logic expressions, a 3-to-8 decoder has a memory of eight stages. The circuit of a 3-to-8 decoder can be implemented using three NOT gates and eight 3-input AND gates as shown in Figure 8.6. An AND gate is used as the basic decoding element for the output because a "HIGH" or logic "1" output is produced only when all of its inputs are logic "1." The three inputs S_0, S_1, S_2 (including their complements) are decoded into eight outputs, D_0 to D_7. A 3-to-8 decoder can be used for decoding any 3-bit code to provide eight outputs, corresponding to eight different combinations of the input code.

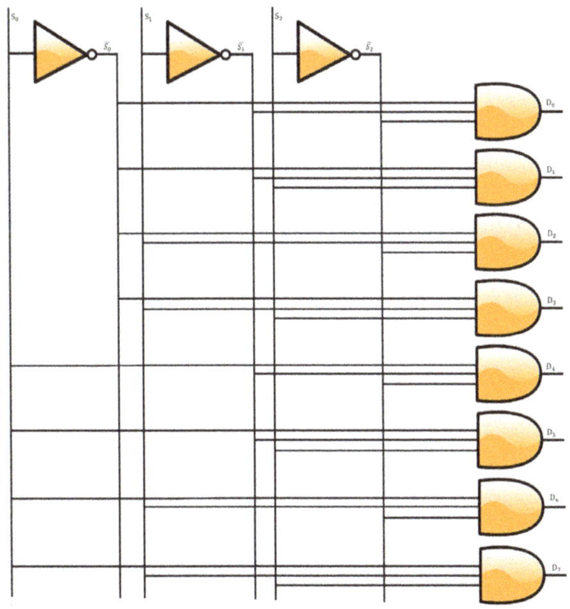

FIGURE 8.6 A 3-to-8 Decoder Logic Diagram

As shown in Figure 8.7 (a) and (b), the physical components in a decoder can be shown or a decoder symbol can be used with the gate details hidden.

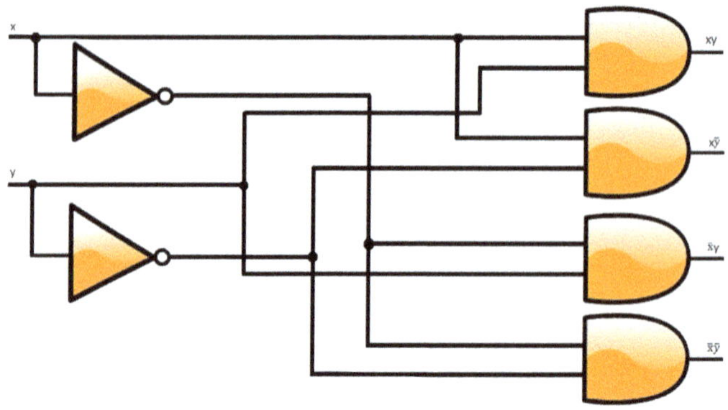

FIGURE 8.7 (A) A Logic Diagram for a Decoder

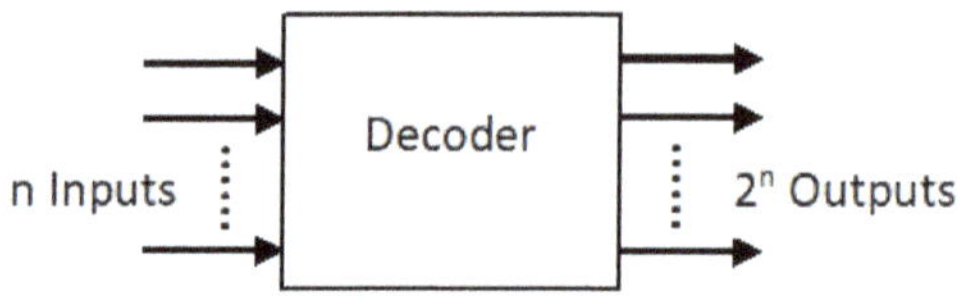

FIGURE 8.7 (B) A Decoder Symbol

As a common combinational circuit, a *multiplexer* is used to select binary information from one of many input lines and then direct it to a single output line. Either control lines or a set of selection variables control the selection of a particular input line. Only one selected input is routed through the circuit to the output line at any given time, and all other inputs are suspended. Any value changes on the control lines cause the route change of the input accordingly. Figure 8.8 (A) and (B) illustrates the logic diagram for a multiplexer and the multiplexer symbol, respectively.

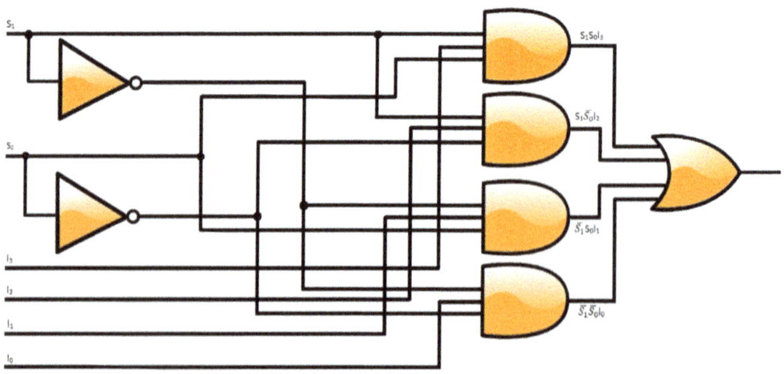

FIGURE 8.8 (A) A Logic Diagram for a Multiplexer

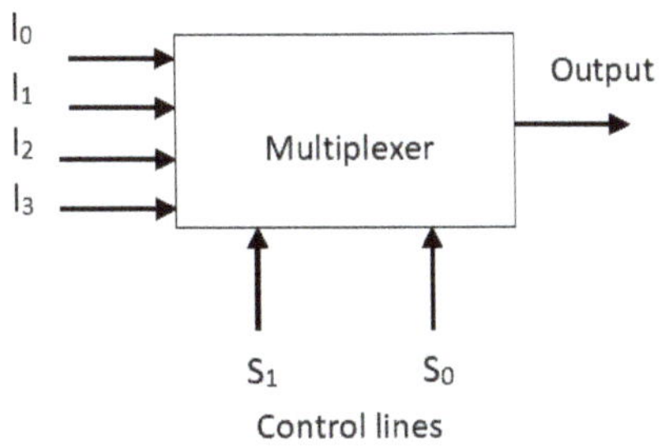

FIGURE 8.8 (B) A Multiplexer Symbol

A parity generator as a combinational logic circuit generates the parity bit in the transmitter. A parity checker as a circuit checks the parity in the receiver. Both parity generators and parity checkers are commonly used in combinational circuits to detect the single-bit errors in the transmitted data.

XOR functions are typically used to construct parity generators and parity checkers. Table 8.5 shows the truth table of a parity generator for a 3-bit input. According to Table 8.5, the expression for P in terms of x, y, and z is given as follows.

$$\begin{aligned} P &= \overline{xyz} + xy\bar{z} + x\bar{y}z + \bar{x}yz \\ &= \bar{z}(x \odot y) + z(x \oplus y) \\ &= x \odot y \odot z \end{aligned}$$

Therefore, an odd parity generator for 3-bit inputs can be built by using two XNOR gates as shown in Figure 8.9.

TABLE 8.5 An Odd Parity Generator for 3-bit Inputs

x	*y*	*z*	PARITY BIT (P)
0	0	0	1
0	0	1	0
0	1	0	0
0	1	1	1
1	0	0	0
1	0	1	1
1	1	0	1
1	1	1	0

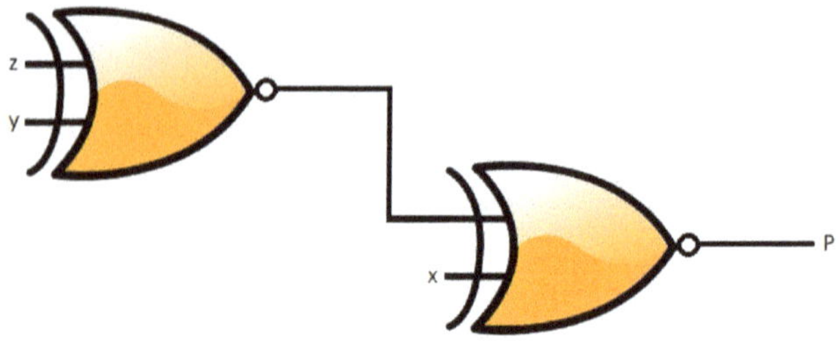

FIGURE 8.9 An Odd Parity Generator for 3-bit Inputs

Table 8.6 shows the truth table of a parity checker for a 4-bit input with 3 information bits and 1 parity. If an error is detected, the parity checker outputs 1; otherwise, output is 0.

TABLE 8.6 Parity Checker

X	*Y*	*Z*	PARITY BIT (P)	ERROR DETECTED
0	0	0	0	1
0	0	0	1	0
0	0	1	0	0
0	0	1	1	1
0	1	0	0	0
0	1	0	1	1
0	1	1	0	0
0	1	1	1	0
1	0	0	0	0
1	0	0	1	1
1	0	1	0	1
1	0	1	1	0
1	1	0	0	1
1	1	0	1	0
1	1	1	0	0
1	1	1	1	1

Sequential Circuits

As discussed in combinational logic, if any variable value changes in Boolean functions, the corresponding function outputs may change accordingly. It depends on whether the variable value change directly affects the outputs. There is no memory storage in the combinatorial circuits. Actually, computers need to save the intermediate values during calculation. How do we design circuits to meet the memory storage requirements of computers? Sequential logic is introduced.

Basic Concepts

A sequential circuit is a logical circuit, where the output depends on the present value of the input signal as well as the sequence of past inputs, while a combinational circuit is a function of present input only. A sequential circuit is a combination of a combinational circuit and a storage element. A flip-flop is used to represent the storage element. A function of the previous inputs to the circuit is the state of this flip-flop. Both the current inputs and the current state of the circuit (the saved previous inputs) work together to determine the outputs.

Clocks

As discussed, the previous inputs are used in a sequential circuit to determine present outputs. Which order should be followed during processing the previous inputs and the current inputs? There are two types of sequential circuits: asynchronous and synchronous. Asynchronous circuits start to process until a certain condition is reached. *Synchronous* ones use clocks to order events. A clock emits a series of pulses with a precise pulse width and a precise interval (called the *clock cycle time*) between consecutive pulses to synchronize the sequential circuits. Megahertz (MHz), or millions of pulses per second, is generally used to measure clock speed. The range of common cycle time varies from one to several hundred MHz.

Most sequential circuits are edge triggered. As shown in Figure 8.10, they are allowed to change their states on either the rising or falling edge of the clock signal.

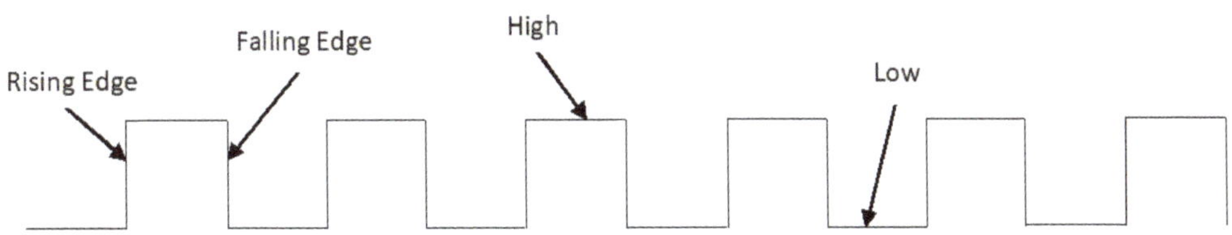

FIGURE 8.10 A Clock Signal

Flip-Flops

A level-triggered circuit is allowed to change state whenever the clock signal is either high or low. A latch is level triggered, and a flip-flop is edge triggered. The term *flip-flop* is discussed here.

The concept of feedback is introduced to sequential circuits to save the previous state. The output of a circuit is fed back as an input to the same circuit. Figure 8.11 shows a diagram using two NOT gates to implement a very simple feedback circuit. If Q is 0, it will always be 0. If Q is 1, it will always be 1.

FIGURE 8.11 A Simple Feedback Example

A basic memory unit SR (short for set/reset) flip-flop composed of two NOR gates is a useful feedback circuit, which is commonly used in memory circuits to store a single data bit. The feedback is provided from both of its outputs back to its opposing inputs. The SR flip-flop has three inputs: set, reset, and its current output relating to its current state or history. Figure 8.12 shows the symbol for an SR flip-flop.

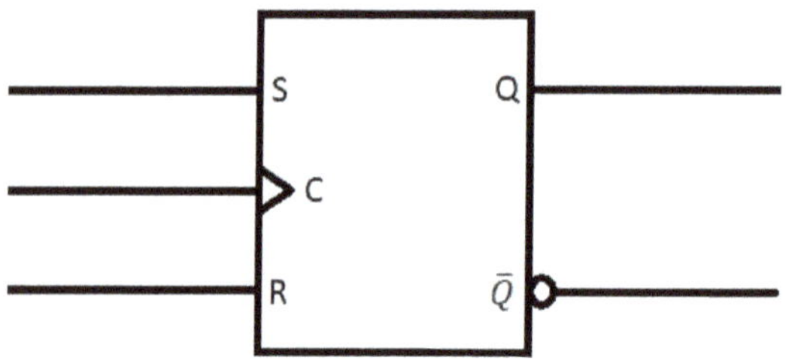

FIGURE 8.12 A Symbol for an SR Flip-Flop

As shown in Table 8.7, a characteristic table is introduced to describe an SR flip-flop. In an SR flip-flop, the next state is based on the inputs and the current state Q.

$Q(t)$ and $Q(t + 1)$ indicate the current state and the next state, respectively. After the clock is pulsed, the state of the SR flip-flop is updated. Figure 8.13 shows the actual implementation of an SR flip-flop.

TABLE 8.7 The Characteristic Table for the SR Flip-Flop

S	R	Q(*t* + 1)
0	0	*Q*(*t*) (no change)
0	1	0 (reset to 0)
1	0	1 (set to 1)
1	1	Undefined

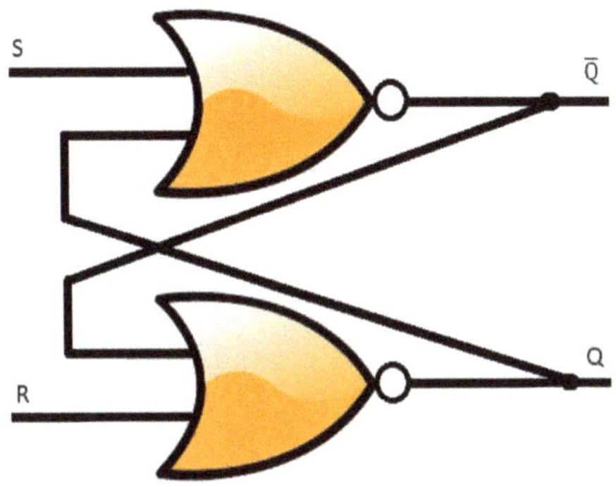

FIGURE 8.13 The Logic Diagram for an Actual SR Flip-Flop

As shown in Table 8.8, the truth table illustrates how an SR flip-flop with three inputs—S, R, and the current output *Q*(*t*)—works.

TABLE 8.8 Truth Table for the SR Flip-Flop

S	R	PRESENT STATE Q(*t*)	NEXT STATE Q(*t* + 1)
0	0	0	0
0	0	1	1
0	1	0	0
0	1	1	0
1	0	0	1
1	0	1	1
1	1	0	Undefined
1	1	1	Undefined

For instance, if S, R, and the current state $Q(t)$ are 0, the next state $Q(t + 1)$ is also 0. If both S and R are 0 and $Q(t)$ is 1, $Q(t + 1)$ is 1. If the input of (S, R) is (0, 0), the next state $Q(t + 1)$ does not have any change when the clock is pulsed. When the input of (S, R) is (0, 1), the next state $Q(t +1)$ is forced to be reset to 0 regardless of the current state. When the input of (S, R) is (1, 0), the next state $Q(t + 1)$ is *set* to 1.

This particular flip-flop has an oddity. When the input of (S, R) is (1, 1), both Q and $\overline{Q}$ are forced to be reset to 1. However, $Q = 1 = \overline{Q}$ is not reasonable. This result causes an unstable circuit. Therefore, the input (1, 1) of (S, R) is not allowed in an SR flip-flop.

As shown in Figure 8.14, a conditioning logic is added to the SR flip-flop to ensure that the illegal state never arises. This circuit is called a *JK flip-flop*. JK is short for Jack Kilby, who invented the integrated circuit in 1958. Table 8.9 shows the characteristic table for the JK flip-flop. Figure 8.15 shows a JK flip-flop as a modified SR flip-flop.

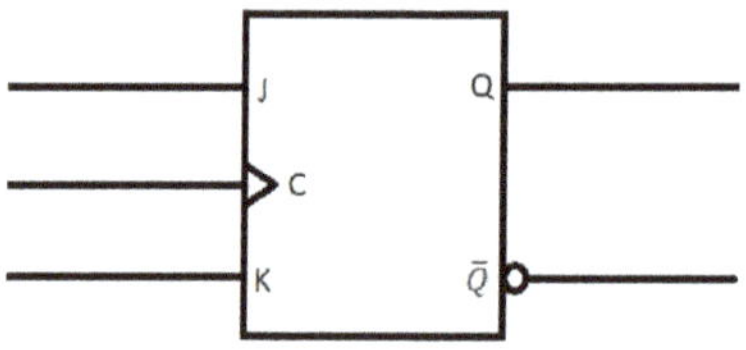

FIGURE 8.14 A JK Flip-Flop

TABLE 8.9 The Characteristic Table for the JK Flip-Flop

J	K	Q(t + 1)
0	0	Q(t) (no change)
0	1	0 (reset to 0)
1	0	1 (set to 1)
1	1	$\overline{Q}(t)$

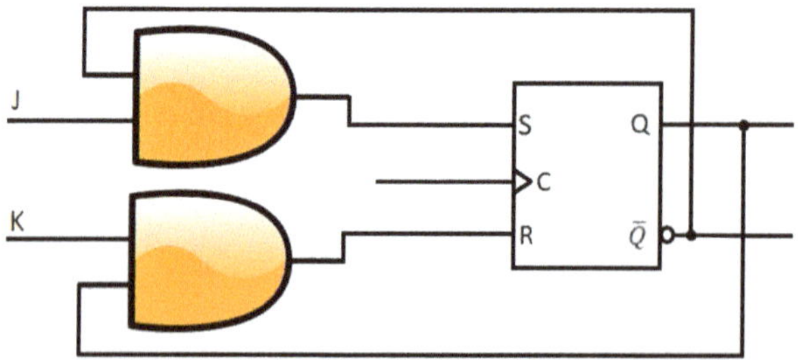

FIGURE 8.15 A JK Flip-Flop as a Modified SR Flip-Flop

As a variant of the SR flip-flop shown in Figure 8.16, a *D (data) flip-flop* is a true representation of physical computer memory. One bit of information is stored in this sequential circuit. Table 8.10 shows the characteristic table for the D flip-flop. If the input of D is 1 and the clock is pulsed, the output of Q is 1. If the input of D is 0 and the clock is pulsed, the output of Q is 0.

The current state of the circuit is reflected in the output Q. For instance, an output value of 1 means the circuit is currently "storing" a value of 1. Figure 8.17 shows a D flip-flop as a modified SR flip-flop.

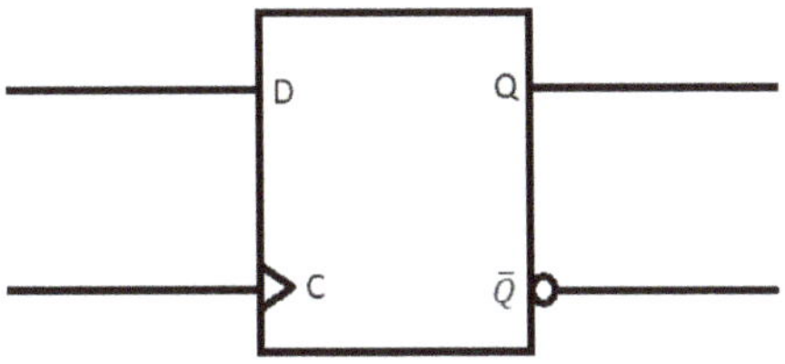

FIGURE 8.16 The Logic Diagram of a D Flip-Flop

TABLE 8.10 The Characteristic Table for the D Flip-Flop

D	Q(*t* + 1)
0	0
1	1

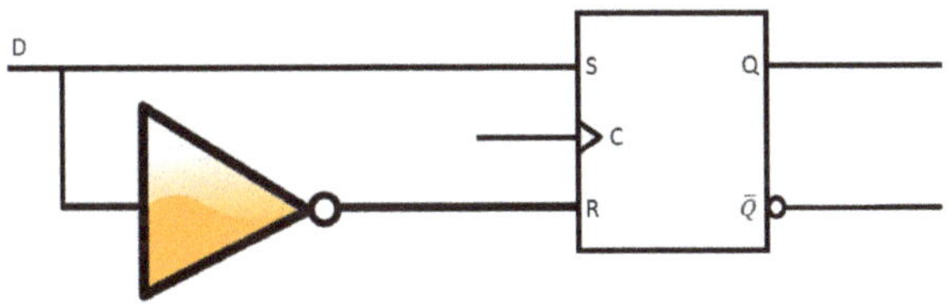

FIGURE 8.17 A D Flip-Flop as a Modified SR Flip-Flop

Chapter Summary

This chapter discusses the definition and the basic design of combinatorial circuits and sequential circuits. The usage of combinational circuits and sequential circuits is covered. The definitions of adders, decoders, multiplexers, parity generators, parity checks, clocks, and flip-flops are covered. The implementation of adder, decoder, multiplexer, parity generator, parity check, and flip-flop is explained in detail and illustrated by using logic diagrams.

Students should master the design and implementation of adder, decoder, and multiplexer, clock, and flip-flop and also be able to apply them to understanding relatively complex designs of combinatorial circuits and sequential circuits.

Chapter 8 Exercises

1. Draw a full adder gate diagram. Label the input and output of each gate for adding 1 + 0 with a carry-in of 1.

2. Draw a gate diagram of a half-adder rippled with a full-adder. Label the input and output of each gate computing the following sum:

$$\begin{array}{r} 1\,1 \\ \underline{+\ 0\,1} \end{array}$$

3. Draw the 3-to-8 decoder shown in Figure 8.6 in the text. Show the input and output of each gate with $S_0 = 0$, $S_1 = 0$, and $S_2 = 1$.

4. Explain how you can predict the output D gate that will have a value of 1 in the 3-to-8 decoder given values of S_0, S_1, and S_2 without having to follow the logic diagram.

5. Consider the multiplexer in Figure 8.8(A) in the text. Fill in the multiplexer output for each of the rows of the following truth table.

I_3	I_2	I_1	I_0	S_1	S_0	OUTPUT
1	0	0	1	0	0	
0	0	1	1	0	1	
0	1	1	0	0	1	
1	0	1	0	1	0	
0	1	1	0	1	0	
1	1	0	0	1	1	

6. Consider the SR flip-flop from Figure 8.13 from your book. Draw the SR flip-flop and show the input and output from each gate for the following values of S, R, and Q(t). This will demonstrate how the flip-flop works.

   a. S = 0, R = 0, Q(t) = 0
   b. S = 0, R = 1, Q(t) = 1
   c. S = 1, R = 0, Q(t) = 0
   d. S = 1, R = 0, Q(t) = 1

7. Explain how the D flip-flop shown in Figure 8.17 prevents an undefined output from occurring.

CHAPTER

9

Computer Memory

Introduction

The concept of storing programs and data in a mechanized system moved non-human computation from simple calculating machines to the basic structure of modern computers. John von Neumann (1903–1957) was a pioneer in mathematics and computer science. He is credited with developing the Von Neumann computer architecture whose components are the basis of computer systems.

The general structure of the Von Neumann architecture is shown in Figure 9.1. Data is communicated to the system from the user through the input device. Data is communicated from the system to the user through the output device. The memory unit acts as long-term storage for data and code. The main job of the central processing unit (CPU) is executing commands. The control unit repeatedly fetches instructions from the memory unit and decodes and executes

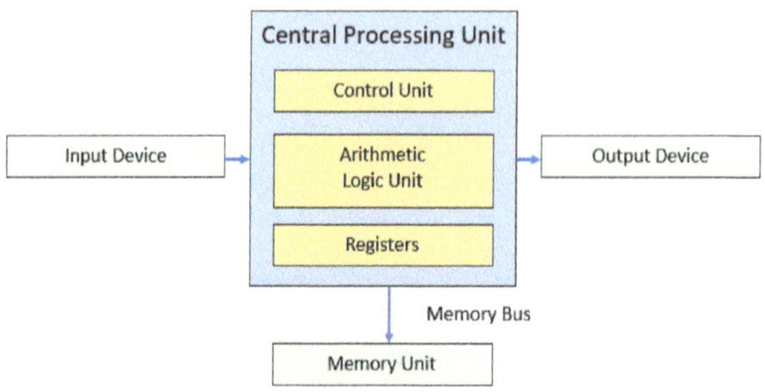

FIGURE 9.1 The Von Neumann Architecture

them (fetch-decode-execute cycle). The arithmetic logic unit (ALU) performs arithmetic and logical operations necessary to process data. Registers are used for fast, local data storage in the CPU. A memory bus is used to transfer data between memory and the CPU. These components and their functions will be described in detail in this chapter.

Even though the Von Neumann architecture is foundational even in modern computers, there have been many improvements made to increase the efficiency of computer systems which are considered non–Von Neumann. The classic Von Neumann architecture has a single processing unit with a fetch-execute cycle. Parallel processing can increase computational speed drastically. Parallel architectures can have many different forms, but in general, multiple CPUs are introduced to perform computation at the same time (in parallel). A problem with the basic architecture is the Von Neumann bottleneck caused by a single bus between memory and the CPU. It is often desirable to fetch multiple instructions and/or data items simultaneously. This becomes an even larger problem when parallel processing is introduced. Non–Von Neumann architectures will have multiple bus configurations to alleviate this problem. It has also become advantageous to include multiple memory units of differing technologies and sizes as opposed to a single memory unit as indicated by the Von Neumann architecture.

A computer system is made up of components designed to process instructions to perform user-defined tasks. The lowest level foundation of the system is digital logic that processes binary data (discussed in the previous units). The major components of a computer system are memory, which allows for the storage of data; the central processing unit (CPU), which processes this data to perform tasks; and the input/output (I/O) system, which communicates data between the computer system and the user. This chapter will concentrate on computer memory. In particular, the chapter will explore the different types of memory in the computer and their uses, hardware used for storing values, and the connection with other components in the computer.

Modern computers have a hierarchy of memory of different sizes and technologies. This hierarchy is designed to ensure speed and cost efficiency in the system. There is a trade-off between speed and dollar cost of memory technologies. The general strategy for designing the memory system is to use large, cheap memories for data that is not currently being used or will not be used in the near future and use smaller, more expensive memories for data that is currently in use.

There are some general principles for all memories, regardless of technology:

- Data is stored in binary (all 1's and 0's).
- Each spot in memory has an address associated with it (like the address of a house) so that the system can find the data when needed.
- There is a cost, measured in time, for accessing data (either reading or writing) in memory.

Memory can be classified into three general categories as shown in Figure 9.2. These memory types will be discussed in the following sections.

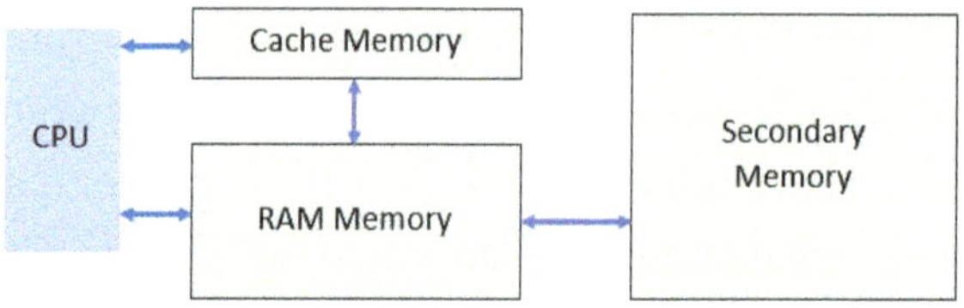

FIGURE 9.2 General Memory Types in a Computer System

Objectives

This chapter will help students:

- **recognize** the Von Neumann architecture as an early design of a computer system.
- **understand** the need for different types of memory in a computer system.
- **illustrate** the relationship between the size of memory and the sizes of addresses.
- **correlate** an address with a location in interleaved memory.
- **articulate** the methods different types of memories use for storing values.
- **understand** the use of random access memory (RAM) in a computer system.
- **explain** how cache memory affects the efficiency of a computer system.

Key Terms

John von Neumann: pioneer in the early development of computers

random access memory (RAM): memory storage for data and instructions in currently executing programs

memory address: a value determining a specific location in memory

cache memory: a small, fast memory unit used to speed execution of programs

secondary memory: memory for long-term storage of programs and data

hard disk drive (HDD): secondary storage using a disk with magnetic medium for storage

solid-state drive (SSD): secondary storage using electronic medium for storage

virtual memory: an extension of main memory to utilize secondary memory

read-only memory (ROM): hard-wired memory for system use that can only be read by the user

Programmable Read-Only Memory (PROM): programmable memory for system use that can be written only one time during system development and can only be read by the user

Erasable Programmable Read-Only Memory (EPROM): programmable memory for system use that can be completely erased and rewritten multiple times during system development and can only be read by the user

Random Access Memory (RAM)

Random access memory (RAM) is a silicon chip memory that is representative of the classic Von Neumann memory unit. RAM or main memory stores instructions and data associated with currently running programs. As the name implies, a random address in memory can be accessed directly. Most commonly, RAMs are separated into bytes (8-bit packages). Each byte in memory has an address that can be accessed. The size of these memories on current PCs are in the gigabyte (GB)—billions of bytes—range, commonly from 2 to 32 GB.

A sequential circuit called a **flip-flop** is used to store each bit of a byte in RAM. Figure 9.3 shows a logic diagram for a flip-flop. Unlike combinational logic, for example, a decoder, a sequential circuit maintains a current state value. In this case, the current state is the value of the bit Q, 1, or 0, stored in memory. For the memory to be useful, the flip-flop's state value must be able to be both read and set.

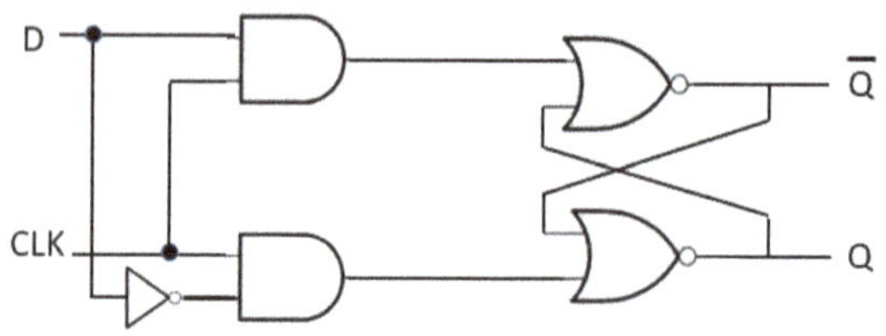

FIGURE 9.3 Flip-Flop Logic Diagram

The two gates to the right are NOR gates, which output the inverse of an OR gate. There are two inputs to the circuit: D, which acts to set the value of the bit, and CLK, which controls whether a change can be made to the value of the flip-flop. The outputs are Q, the value of the flip-flop, and NOT Q, the inverse of the value of the flip-flop. The current value of Q allows the system to read a bit value at a certain address. The value of Q is retained based on the feedback loop represented by the values of Q and NOT Q being sent back into the two NOR gates at the right of the figure. If no new inputs are sent to a flip-flop, the feedback loop will cause the flip-flop to be in a stable state (unchanging).

The values of the inputs D and CLK allow the system to set the value of Q in the flip-flop. The value of D represents the value to assign to Q. However, we will see that a value of 0 for CLK will not allow the value of Q to change. Table 9.1 shows the value of the flip-flop based on the current state, current value of Q, and the inputs from D and clock. The result is Q_{T+1} (the new value of Q at time T + 1).

TABLE 9.1 Flip-Flop Truth Table

D	CLK	Q_T	Q_{T+1}	RESULT
0	0	0	0	Q remains 0
0	0	1	1	Q remains 1
0	1	0	0	Q remains 0
0	1	1	0	Q is set to 0
1	0	0	0	Q remains 0
1	0	1	1	Q remains 1
1	1	0	1	Q is set to 1
1	1	1	1	Q remains 1

Example 9.1 demonstrates that the value of the flip-flop can change when CLK is equal to 1. Refer to Figure 9.4 illustrating the situation. The red bits are initial values that are changed to final blue bit values in this case.

Example 9.1: Row 4 of the table setting the value of Q to 0.

D = 0 (setting value of flip-flop to 0); CLK = 1 (allows the flip-flop to change value)
Q = 1 (current value of flip-flop is 1)
Steps in setting Q to 0 and reaching the stable state:

1. Output from upper AND gate is 0 (input D is 0, input CLK is 1).
2. Output from lower AND gate is 1 (input NOT D is 1, input CLK is 1).
3. Output from lower NOR gate is 0 (input of 1 and feedback of 0 from NOT Q)

Note: The value of the flip-flop, Q, has changed.

4. Output from upper NOR gate is 1 (input of 0 and feedback of 1 from Q).
5. Output from lower NOR gate remains 0 (input of 1 and feedback of 1 from NOT Q).

Flip-flop is now in a steady state of 0, Q at time T + 1.

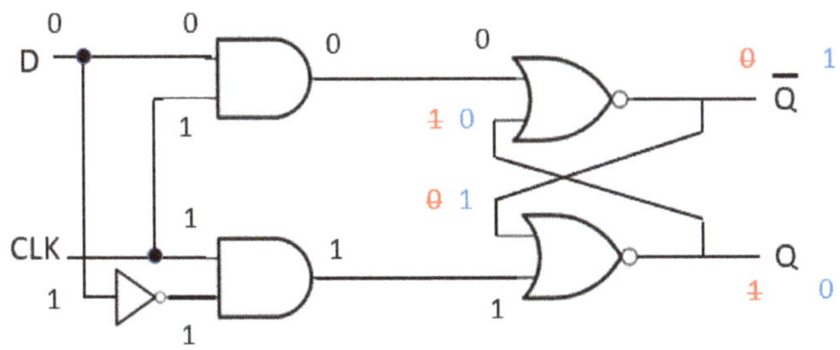

FIGURE 9.4 A Value of CLK = 1 Allows the Value of the Flip-Flop to Change to 0

Example 9.2 demonstrates that the value of the flip-flop does not change when CLK is equal to 0. (Refer to Figure 9.5 illustrating the situation.) The output blue bit values exiting the flip-flop do not change.

Thus, the value of the flip-flop can only be changed when the CLK is equal to 1. This value of CLK could be the value of the system clock, which would allow the system to only change the value of a bit at an address at certain times. CLK

Example 9.2: Row 2 of the table with the value Q not changing.

D = 0 (might set value of flip-flop to 0); CLK = 0 (prevents flip-flop value change to 0)

Q = 1 (current value of flip-flop is 1)

Steps in reaching the steady state:

1. Output from upper AND gate is 0 (input D is 0, input CLK is 0).
2. Output from lower AND gate is 0 (input NOT D is 1, input CLK is 0).
3. Output from lower NOR gate is 1 (input of 0 and feedback of 0 from NOT Q).

Note: The value of the flip-flop, Q, does not change.

4. Output from upper NOR gate is 0 (input of 0 and feedback of 1 from Q).
5. Output from lower NOR gate remains 1 (input of 0 and feedback of 0 from NOT Q).

Flip-flop remains in a steady state of 0 (Q and NOT Q have not changed).

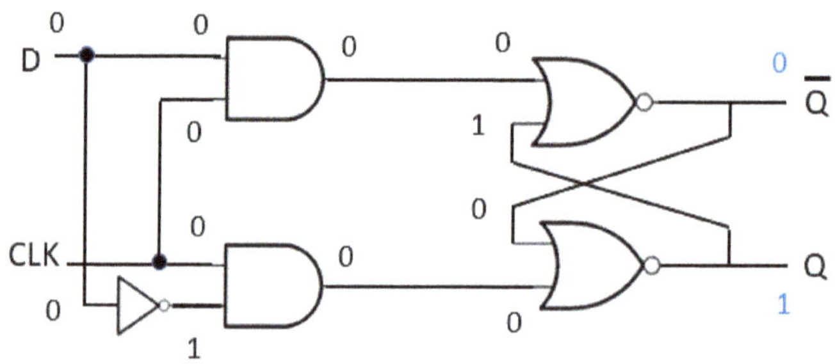

FIGURE 9.5 A Value of CLK = 0 Does Not Allow the Value of the Flip-Flop to Change

may also be 1 if a bit address is chosen. Flip-flops in other addresses would be sent a value of 0 for CLK and not change.

The speed of access of RAM memory is measured in **megahertz (MHz),** which represents millions of cycles per second. A typical speed of access would be 1000–2000 megahertz. Given a 1000 megahertz memory access rate, a single memory access would take 1 **nanosecond (ns)**, a billionth of a second. Currently running programs and data are loaded into RAM to achieve this fast access. RAM sizes often run from 2 GB (billions of bytes) to 32 GB. Larger RAM sizes are better for the efficient running of programs. It is undesirable to have program code on slower secondary storage devices.

Memory Addresses

A byte-addressable memory will allow a program to access each individual byte in RAM. Each byte in memory must have a unique address. The size of memory in bytes will dictate the number of bits in an address; *n* bits can address of memory of 2^n bytes. This was shown in Chapter 1 where, for example, we can represent 2^3 or 8 possible values using three bits (values 0 through 7 in base 10). Therefore, a memory of 4 GB would have addresses from 0 to 4,294,967,295. This is 2^{32}, which is equal to 4,294,967,296. The 32-bit address at the start of memory would be 0 (32 bits of value 0) and the address at the end of memory would be 4,294,967,295 (32 bits of value 1). Each addressable byte in memory will be made up of eight flip-flops.

As seen in the last chapter, a decoder is able to select single output line to have a value of 1 (essentially activating that distinct output). Therefore, a decoder is used to select a particular byte in memory based on its address. For the memory described above, this is a 32-to-4,294,967,295 decoder! That is, the 32-bit input address will select one of the 4,294,967,295 bytes in memory.

Not all machines are byte addressable. Some have a larger word size that is addressable. Perhaps each word would be comprised of 4 bytes. In this case, the number of address bits would be 2 less for the same number of bytes as a byte addressable. Figure 9.6 shows a simple example. A 32-byte byte-addressable memory would need 5 address bits ($2^5 = 32$). The same 32-byte memory with 4-byte words would have 8 addressable words and only need 3 address bits ($2^3 = 8$).

Address	Byte 1	Byte 2	Byte 3	Byte 4
000	0	1	2	3
001	4	5	6	7
010	8	9	10	11
011	12	13	14	15
100	16	17	18	19
101	20	21	22	23
110	24	25	26	27
111	28	29	30	31

FIGURE 9.6 A 4-byte Word-Addressable Memory

For manufacturing reasons, it is common to have main memory stored on more than one chip. For example, a 32-MB memory might be made up of eight 4-MB chips. This structure is called an **interleaved memory**. These physical chips are logically called modules. The advantage of interleaved memory is that blocks of data from different modules can be accessed at the same time achieving parallelism. This can greatly improve the access time of main memory. To find an address in memory, it is now necessary to specify both the module being accessed and the address of the data in the module. This information is given as one address, with the high-order bits representing the module and the low-order bits representing the address of the data within the module.

As a simplified example, consider Figure 9.7 which has 32 words in memory separated into four modules. Each module consists of eight words. The address of each word is shown in the figure. To access a particular word, the high-order 2 bits determine the correct module (selecting one of the four modules). The low-order 3 bits select the correct word within the module (selecting one of the eight words).

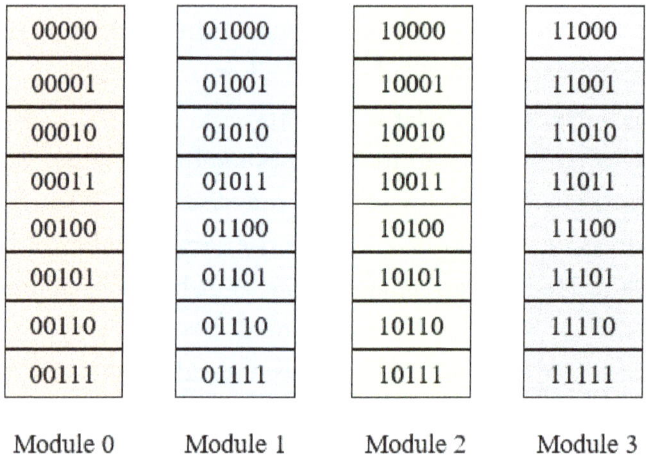

FIGURE 9.7 Interleaved Memory with Four Modules and Eight Words in Each Module

Example 9.3: Referencing a word in the interleaved memory in Figure 9.7.

- In what module would the word at address 14 be found?

  14 in 5-bit binary: 01110
  High-order 2 bits give the module: 01110
  01 ➔ Module 1

- What is the index of the word in Module 1 where address 14 will be found?

  14 in 5-bit binary: 01110
  Low-order 3 bits give the index: 01110
  110 ➔ Index 6

Cache Memory

Fast memory access is important for the efficient running of programs since memory access is frequently needed and slower than CPU processing. Programs may be run faster if the speed of memory access is increased. **Cache** memories are smaller and faster than RAM. These memories will often reside on the CPU chip. Cache memory sizes are typically in the megabyte range (millions of bytes), so they are a factor of 1000 smaller than RAM memories, which are in the gigabyte range. These smaller memories have access 10 to 100 times faster than RAM speeds. Many computer systems use several cache memories of varying sizes to further increase the speed of access.

The key to success of the use of these smaller memories is the movement of frequently accessed blocks of instructions from RAM to cache. When a memory value is needed, the cache is accessed first to see if the value resides there. When a memory access is found in the cache, it is called a **hit**, while a memory access not found in the cache is called a **miss**. Consider the following simplified formula for computing the average access speed using a cache:

$$\text{Average Access Speed} = C_{access} + (1 - \text{Hit}) \times R_{access}$$

where C_{access} is the time to access a byte in the cache, R_{access} is the time to access a byte in RAM, and Hit is the ratio of times that a byte is found in the cache. The first term, cost of cache access, represents the idea that the cache is always accessed first. The second term, cost of RAM access, is only added when there is a cache miss.

Example 9.4: Computing average access speed in a cache.

Compute the average access speed with $C_{access} = 5$ ns, $R_{access} = 70$ ns, and Hit = 0.7:

$$\text{Average Access Speed} = 5 + (1 - 0.8) \times 70 = 19 \text{ ns}$$

Compute the average access speed with $C_{access} = 8$ ns, $R_{access} = 80$ ns, and Hit = 0.9:

$$\text{Average Access Speed} = 8 + (1 - 0.9) \times 80 = 16 \text{ ns}$$

As seen in the examples, there is an improvement using the cache over the 70 or 80 nanoseconds average access time if the cache was not present. Note that a small hit ratio can cause the cache system to have a higher average access speed. The typical success of the use of a cache is aided by the general idea that programs often execute code in small blocks in their programs due to sequential execution and looping structures. This is called the **locality principle**. This execution of programs using a small area of memory which can fit into cache will generate a high hit ratio and subsequently a fast average access speed.

Secondary Memory

Secondary memory is used for storing code and data (text, images, audio, or video) that is not being currently processed. Unlike RAM and cache, which is volatile (needs power to retain values), these memory technologies are nonvolatile and store data when the computer is powered down. When a program begins execution, it is loaded into RAM from secondary storage.

A common secondary memory is a **hard disk drive (HDD)**. Data is stored on a magnetic medium in concentric tracks. This memory is nonvolatile since power is not needed to maintain the magnetic charge. Transitions in the magnetic charge define the bit values of 1s and 0s. Each track is separated into blocks of data, which are loaded as a unit into RAM. A movable read-write head finds the track of the block that is needed. Figure 9.8 shows a disk drive with read-write head. The disk spins to move to the desired block within the track. Therefore, the access time is defined as the **seek time** (amount of time needed for the read-write head to find the correct track) plus the **rotational latency** (amount of time for the spinning disk to find and read the correct block). These mechanical processes cause disk data access to be considerably slower than RAM access. There are several factors that influence difference in speed of transfer (size of transfer, position of read-write head both

FIGURE 9.8 Hard Disk Drive with Magnetic Disk Storage and Read-Write Head

between tracks and within a track, sequential versus random access) but access of data on RAM is generally a factor of 100 times faster than on a hard drive.

Although the access speed of a hard disk drive is considerably slower than RAM, HDD storage is cheaper, larger, and is nonvolatile so it is a good medium for long-term internal storage on a computer. While RAM memory size is in the gigabyte range, disk drive storage is in the multi-terabyte (TB)—trillions of bytes—range.

A more recent development in secondary storage is the use of **solid-state drives (SSD)**. These are electronic drives that use flash memory technology. Similar to RAM storage, these drives have data that can be randomly accessed, unlike HDD memory, which must be accessed sequentially. The SSD memory is a nonvolatile medium using floating gate transistors that can maintain an electrical charge. SSD memory has advantages over HDD storage, including reliability due to nonmovable parts, a constant speed of access, ability to perform random access, increased speed of access, faster start-up, quieter operation, and no need for a fan to cool the unit. Disadvantages of SSD as compared to HDD include increased cost, smaller size of memory, and degradation of access speed over time. Currently, a large PC SSD drive would be 1 TB.

There are external versions of HDD and SDD using the same technologies allowing for backup storage of data and transfer of data to other computers. These include external USB flash drives, external hard drives, and CD drives.

Virtual Memory

Many modern computer systems are able to treat secondary storage as an extension of RAM or main memory. This process is called **virtual memory**. In virtual memory, a program in secondary storage is loaded into main memory in blocks called pages. RAM consists of a fixed number of page frames to accept the pages being transferred. In a **demand paging** system, pages are only loaded into RAM when needed. If the page referenced in the program is not in RAM, this is called a **page fault,** and the page is loaded at that time. Demand paging can speed the execution of a program, since pages only need to be loaded if used. In general, virtual memory allows programs that are larger than main memory to be run. It also facilitates faster execution of **multiprogramming**, multiple programs that are running at the same time. As with cache memory, virtual memory works best when small sections of the program are executed at a given time. That is, as addresses are referenced, the data is already in main memory, giving a high hit ratio.

As memory fills, there are times at which sections of code must be removed from main memory to make room for code that has just started execution. Different algorithms have been developed to select which page resident in main memory will be removed (the **victim**). The goal is to remove a page of code which will not be used in the near future or ideally not used at all. Two common methods for selecting a page to remove are first-in first-out (FIFO) and least recently used (LRU). FIFO will remove the page that was loaded into main memory the earliest. LRU removes the page that has not been accessed recently. Both of these algorithms are trying to predict which page is no longer needed.

Example 9.5 demonstrates the difference between FIFO and LRU page replacement algorithms.

Example 9.5: FIFO and LRU page replacement.

Assume that there are four page frames available in RAM. The following virtual memory page references occur during the running a program: 1, 5, 7, 2 (note at this point four pages have filled the four page frames), 7, 1, 8 (no room for page 8, need to select a victim).

FIFO: Removes page 1 (first of the four pages to enter RAM)

LRU: Removes page 5: (least recently of the four pages to be referenced.

ROM, PROM, and EPROM

There is code necessary to boot your computer on start up. This memory naturally needs to be nonvolatile. **ROM (Read-Only Memory)** is a class of memory which is often used for this purpose. As the name suggests, ROM can only be read but not written to after manufacture. ROM is designed specifically for a particular computer architecture. If there is software added which was not handled by the original architecture or errors are detected, the ROM unit must be replaced. ROM is hardwired using digital logic. The cost of developing ROM is high as the circuit must be designed and machinery adapted to print the circuit. Typically, this process is expensive and is only worthwhile when large quantities are needed. As mentioned above, if problems are encountered on application, the design and production steps must be repeated. ROM, like the other memories discussed in this section, is used in many types of devices where memory, or at least parts of it, remains static—devices such as phones, calculators, car electronic systems, and video games.

Programmable Read-Only Memory (PROM) is a nonvolatile memory that solves some of the problems of ROM as it can be changed once after it has been manufactured.

PROM cannot be changed after that first write. It has the advantage over ROM in that it is not designed as digital logic but is microprogrammed. This decreases the production time and allows for smaller numbers of units to be created for a particular machine. Errors in the design can be more easily adjusted as program changes are made as opposed to redesigning and printing hardware. However, once a PROM has been written to, it cannot be changed. Therefore, on failure or redesign, a new unit must be manufactured.

Erasable Programmable Read-Only Memory (EPROM) is a nonvolatile memory that has the advantage over PROM in that it can be written multiple times. When EPROM is written to again, the entire memory is erased, using a process that exposes it to ultraviolet light. Figure 9.9 shows an EPROM chip. Note that the memory itself is exposed in the middle of the container, allowing access to ultraviolet light for clearing the data. This makes EPROM an excellent medium for development of computer systems as it can be reprogrammed. Once debugged, the developed EPROM memory can be translated to a hardwired memory. EPROM technology is similar to that of SSD and flash memory using floating gate transistors, but EPROM is a smaller, faster, and more expensive medium.

FIGURE 9.9 EPROM Chip with Exposure to Ultraviolet Light

Chapter Summary

The need for memory storage was recognized as being an essential part of computer systems from their earliest designs. The use of binary allowed computers to develop technologies to store all types of data with speed and reliability. One of the important attributes of the classic Von Neumann computer architecture design included storage of program code in memory, greatly enhancing the efficiency of program execution.

The use of a hierarchy of different memory technologies optimizes a computer system in terms of execution speed and cost. Main memory or RAM is the basic component for storing data and instructions of a currently running program. RAM is composed of volatile memory components called flip-flops, each of which stores one bit of data. The bits are grouped together into larger addressable units, often 8-bit units called bytes. In this case, a byte can be quickly and randomly accessed from memory to be loaded into the CPU for processing. Often, computer memories are separated into multiple byte units called words. An address specifies the position of a unique word in memory. Memory can be subdivided into modules using memory interleaving, where addresses specify both a module and a word in that module.

Small, fast, and expensive cache memories are utilized in a computer system to speed processing. Frequently accessed program code is loaded into cache to take advantage of the greater speed of access. This greatly improves the efficiency of execution if a program has significant locality of reference generating a high cache hit ratio.

High-volume, secondary memory devices such as HDD and SSD are less expensive. These devices are nonvolatile and thus can be used for long-term storage of data. Virtual memory extends RAM into these secondary storage devices to optimize the use of memory for large programs and multiprogramming computer systems.

The computer system needs program code to start execution when the computer is powered on. There is a need for nonvolatile memory to store this boot code. There are a variety of technologies that are used to store this code. ROM uses digital logic to store this memory. PROM allows memory to be written to once, using microcode. EPROM allows multiple writes to the memory. To accomplish this, the current memory must be completely erased.

Chapter 9 Exercises

1. Consider the Flip-Flop Logic Diagram shown in Figure 9.3. Show the input and output from each gate for the following examples. If a bit values changes, cross out the old value and add the changed value.

   a. D = 1, CLK = 1, Q(t) = 0
   b. D = 1, CLK = 0, Q(t) = 0

2. How many address bits are needed for each of the following memory sizes?

   a. 4096
   b. 65,536
   c. 67,000,000,000

3. Consider an interleaved memory with 16 modules and 128 words in each module. Note for the following answers, the first module and the first address in a module each have a value of zero.

   a. Give the number of bits needed to reference a word in this memory. Explain your answer.
   b. Give the full address of the word 7 in module 5.
   c. Give the module that the following address will select: 10011100110.
   d. Give the word in the selected module that the following address will select: 10110011101.

4. Compute the average memory access speed given the following values.

   a. cache access speed: 6 ns; RAM access speed: 18 ns; hit ratio: 0.7
   b. cache access speed: 3 ns; RAM access speed: 12 ns; hit ratio: 0.8
   c. cache access speed: 8 ns; RAM access speed: 20 ns; hit ratio: 0.9

5. Consider a hard disk drive memory.

   a. Define seek time.
   b. Define rotational latency.
   c. Assume that you have a number of disk access requests waiting to be processed with track and block numbers for each request available. Explain how these requests might be ordered to increase system speed.

6. Consider a virtual memory system with four page frames with demand paging and the following sequence of page references: 1, 2, 3, 4, 3, 1, 2, 6, 4, 9, 8, 1, 3, 8, 7, 2. Show the contents of the four page frames after each reference using the following algorithms. Additionally, indicate the number of page faults.

   a. FIFO page replacement algorithm
   b. LRU page replacement algorithm

7. Explain why basic ROM would be a preferable memory for use in a calculator as compared to EPROM.

8. Explain why EPROM would be a preferable memory for using in a general purpose computer as compared to basic ROM.

Credits

CHAPTER 10

Central Processing Unit (CPU)

Introduction

The **central processing unit (CPU)** executes programs. Figure 10.1 shows the important components of the CPU.

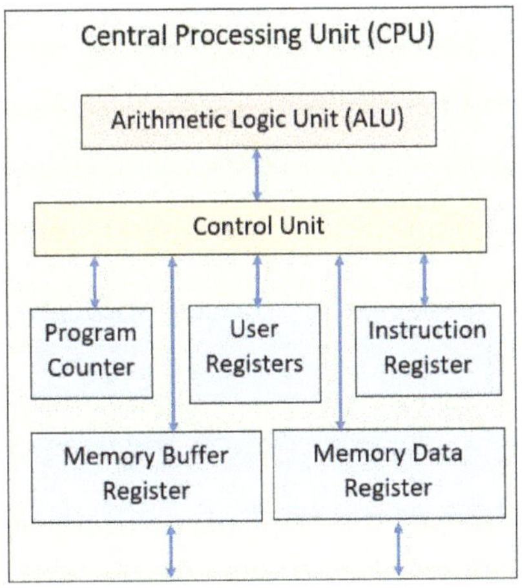

FIGURE 10.1 CPU Components

The **control unit** controls the execution of the program, including reading and writing to memory. The **arithmetic logic unit (ALU)** performs arithmetic and logical computations. CPU registers are memory units with fast access. These registers store data and addresses used during program execution. Two registers often included in the control unit are the **memory address register (MAR)** and **memory buffer register (MBR)**. The MAR stores an address in memory. The MBR stores data associated with a memory access. Thus, these two registers work in tandem to read or write data between the control unit and memory. The **program counter (PC)** stores the memory address of the next program instruction to be executed. The **instruction register (IR)** stores the instruction currently being executed.

Objectives

This chapter will help students:

- **connect** CPU registers with their uses in program execution.
- **understand** the fetch-decode-execute cycle of program execution.
- **explain** the relationship between main memory and the CPU.
- **compare** hardwired control and microprogrammed control.
- **identify** operations performed by the ALU.
- **interpret** digital logic in the ALU.
- **summarize** the different parallel processing paradigms.

Key Terms

central processing unit (CPU): module of a computer system that executes programs

fetch-decode-execute cycle: repetitive process indicating the steps in executing instructions

control unit: portion of the CPU that controls the fetch-decode-execute cycle

program counter (PC): address of the next instruction to be executed

register: fast memory storage unit in the CPU

instruction register (IR): a register storing the instruction being executed

opcode: a binary value indicating the operation to be executed

hardwired control: a control unit that is comprised of digital logic

microprogrammed control: a control unit that uses a microprogram to control execution

memory address register (MAR): register holding the address of a data being transferred to/from memory

memory buffer register (MBR): register holding data being transferred to/from memory

arithmetic logic unit (ALU): component that performs arithmetic and logical operations

pipelining: a processing system made up of a sequence of steps where different steps can be executed in parallel

single instruction/multiple data (SIMD): a synchronous processing system that has multiple processors executing the same instructions in parallel

multiple instruction/multiple data (MIMD): an asynchronous processing system that has multiple processors executing different instructions in parallel

multicore processor: CPU with two or more processing units, each running its instruction execution cycle

graphics processing unit (GPU): parallel processing unit with a large array of core processors designed for fast graphics data processing

Registers

The types of memory discussed so far are not located on the CPU chip and must be retrieved using bus communication, which is a slow operation. Access to memory off of the CPU chip is further undesirable due to congestion on the bus, which causes a wait time for retrieval of data. **Registers** are fast, volatile, on-chip memory storing an individual data item. There are a small number of such registers available due to their expense and the need to minimize chip size.

Registers on the chip fall into certain specialized categories. Some registers are available to the user (programmer) for use. Once again, the idea of locality of reference comes into play. Since registers represent the fastest memory access, there is a goal of using them to store the most frequently used data items. High-level language compilers (such as for C++) will optimize code to accomplish this.

Examples of categories of **user-accessible registers** include:

Data registers: data that can represent different types of values, such as integer, floating point, and character

Address registers: memory addresses used for cataloging positions of data in memory that can be used, for example, as pointers, stack addresses, or indexes into an array

General purpose registers: registers that at a time can represent either data or addresses
Program counter (PC): address of next instruction to be executed
Program status register (PSR): register with bits that specify the state of a result of an instruction such as a carry-out, overflow, parity (odd or even number of 1s), or negative value

The system uses **internal registers** that are not accessible to the programmer. These include:

Memory buffer register (MBR): stores a value written to or read from memory
Memory address register (MAR): the address of memory to be used for a read or write instruction
Instruction register (IR): the current machine language instruction being executed
Output register (OutReg): register holding data to be sent to an output device
Input register (InReg): register receiving data from an input device

Control Unit

The control unit repeatedly executes a fetch-decode-execute cycle to run programs.

Fetch: The PC stores the address of the next instruction, which the control unit fetches and loads into the **instruction register (IR)**. The instruction is binary machine language. Machine language is the native language of a computer (understood implicitly by the hardware). After fetching the instruction, the PC is incremented to address the next instruction. The fetched binary instruction is then decoded. A sample format for an instruction is shown in Figure 10.2.

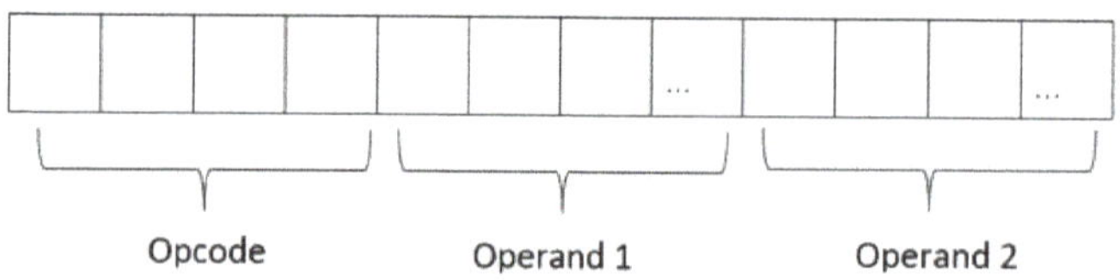

FIGURE 10.2 Sample Instruction Format

Decode: The decode step looks at the opcode portion of the instruction to determine the operation to be performed. The number of bits in the opcode indicate how many possible instructions are available. In Figure 10.2, there are 4 bits, which would allow for 16 (2^{bits}) different instructions. A decoder can be used to select which of the instructions was chosen and activate the code associated with that instruction. Recall that a decoder takes a number of input bits, in our example 4 bits in the opcode, and will have a 1 on the desired output line with 0s on all (in this case, 15) other lines. The chosen output line will lead to the execution of the desired operation specified by the opcode.

Execute: The execute step, depending on the type of instruction, may use the ALU for computation and/or move a value (perhaps the result of a computation) to a register or memory. The first step in execution is to access the operands, if needed for the given instruction. The references to the operands are in the instruction as shown in Figure 10.2. The operands may be found in CPU registers, in memory (in which case the operand will be a memory address), or in the instruction itself. An **addressing mode** specified in the instruction will indicate where the operand is located. The number of bits in the operand, similar to the opcode, control the number of values that can be referenced. Thus, if there is a 16-bit operand being used as an address, then the instruction can address 2^{16} or 656536 memory locations.

If there is a computation performed by the instruction, the resulting value is stored. Depending on the format of the instruction and the architecture of the computer, this value might be stored in one of the two operands given, in a third operand in the instruction, or in a predefined location, such as the top of a stack. There is a **program status register (PSR)** that is set after each instruction with bits that indicate various possible results of the computation such as overflow (a problem with the size of the result), positive result, negative result, zero result, and parity of result (a problem with the bits in the result). The values in the PSR are used in conjunction with conditional branch (go to) instructions (was the result of a computation zero?) or to detect error conditions.

As mentioned above, the address of the next instruction to be executed has been incremented in the PC with the assumption that the next sequential instruction will be used. If the given instruction has a branch associated with it, the control unit must load the branch address into the PC.

There are two general designs for the control unit: hardwired control and microprogrammed control. A **hardwired control unit** is constructed from digital logic components to guide the operation of the fetch-decode-execute cycle. A decoder is used with input from the opcode portion of the instruction to select logic that performs the correct operations. Based on that choice, a sequence of steps is executed to complete the instruction task. These steps typically include fetching needed operands from memory/registers, using the **arithmetic logic unit** (**ALU**) to compute a result and writing this result to the desired memory/register. Memory accesses utilize the MAR and MBR. These steps may be further divided; for example, the ALU may also be needed to compute memory addresses for data transfer. **Microprogrammed control** uses a simple programming language made up of **microinstructions** to control execution of a program. Each microinstruction specifies a microoperation. Some examples of microoperations are reading a value from memory; using the ALU for a computation; incrementing the program counter; or writing an address into the MAR. The microinstructions are located in a special section of **Read-Only Memory** (**ROM**) called the **control store**. A **microprogram counter** (**MPC**) is analogous to the PC, which has the address of the next microinstruction to be executed. Microprogrammed control is designed to be very simple. The code for a given instruction, say an ADD command, is stored in a sequence of instructions in the control store.

The steps for executing the fetch-decode-execute cycle for microprogrammed control are:

1. Fetch the instruction address specified by the PC into the IR (step at MPC = 0)
2. if opcode is ADD

    branch to add code sequence in the control store (set the MPC)

3. if opcode is SUB

    branch to subtract code sequence in the control store (set the MPC)

    ...

4. Execute the sequence of instructions starting at the MPC to process the instruction
5. Set MPC = 0 (back to step 1)

Arithmetic Logic Unit (ALU)

The arithmetic logic unit (ALU) performs arithmetic and logical operations. Depending on the instruction set available in the computer, operations that it can perform include arithmetic operations (addition, subtraction, multiplication, division), logical operations (AND, OR, NOT, EX-OR), and bitwise operations, such as shifting bits left or right in a data word. The ALU has two major types of input. One set of input will be the data to be processed. Another set of input will be control, which indicates the operation that will be applied to the data. Figure 10.3 shows a simple 1-bit ALU structure.

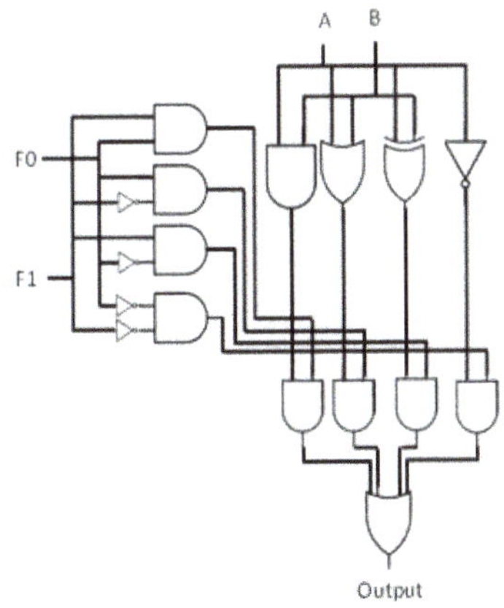

FIGURE 10.3 Simple 1-bit ALU Structure

The two input data bits, A and B, are shown at the top of the diagram. At the left of the figure is a decoder, which chooses one of four logical operations (A AND B, A OR B, A EX-OR B, or NOT A) based on the control bits F0 and F1, which are supplied by the control unit. Note that all four of the logical operations are computed, but the output of only one is chosen by the multiplexer at the bottom of the figure. A truth table showing all possible inputs and associated outputs is shown in Table 10.1.

TABLE 10.1 ALU Truth Table

F1	F0	A	B	OUTPUT	OPERATION
0	0	0	0	1	NOT A
0	0	0	1	1	NOT A
0	0	1	0	0	NOT A
0	0	1	1	0	NOT A
0	1	0	0	0	A OR B
0	1	0	1	1	A OR B
0	1	1	0	1	A OR B
0	1	1	1	1	A OR B
1	0	0	0	0	A X-OR B
1	0	0	1	1	A X-OR B
1	0	1	0	1	A X-OR B
1	0	1	1	0	A X-OR B
1	1	0	0	0	A AND B
1	1	0	1	0	A AND B
1	1	1	0	0	A AND B
1	1	1	1	1	A AND B

Example 10.1: Computing an output using the ALU–Row 6 of Table 10.1.

Inputs: F1 = 0; F0 = 1; A = 0; B = 1
Following the logic gates:

1. Decoder AND gate outputs at left of the circuit are (top to bottom): 0, 1, 0, 0.

   The AND gate with inputs NOT F1 and F0 has two 1 inputs and thus a 1 output.

2. Inputs to the AND gates at the bottom of the circuit (multiplexer) from the decoder: 0, 1, 0, 0.

   The three AND gates with 0 inputs must have 0 outputs.
   The second AND gate has an input of 1 and the output of A OR B from the gate above.
   Note that when F1 = 0, F0 = 1, the OR operation is always chosen.

3. The value of A OR B is: 0 OR 1 ➔ 1.
4. The inputs to the four input OR gate at the bottom will thus be: 0, 1, 0, 0.

   The Output will be 1.
   Conclusion: Given the inputs in the example, the Output of the ALU is:

   A OR B ➔ 0 OR 1 ➔ 1

   as indicated in row 6 of Table 10.1.

A typical 1-bit ALU will perform many more operations (will have more control bits) and will also have a carry-out bit for operations like addition that can generate a carry. Of course, there is a need to have an ALU, which works on data with many bits as opposed to just one. One-bit ALUs can be rippled to work on a larger number of bits. Figure 10.4 shows an example of a 4-bit ripple-carry ALU.

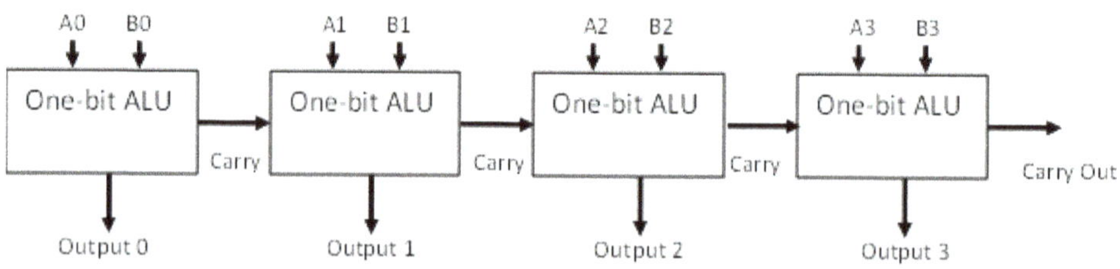

FIGURE 10.4 Four-bit Ripple-Carry ALU

Parallel Processing

The classic Von Neumann architecture prescribes a single processing unit. **Moore's law** states that the number of transistors on chips doubles approximately every 18 months. Advances in computer technology have followed this trend for decades, leading to smaller and faster processors. An Intel Quad Core processor has over 1 billion transistors in a small package (Figure 10.5). However, many CPU bound algorithms such as those pertaining to big data, searching, and neural networks have put an even greater emphasis on speed of processing. To fill this need, computers have been designed to implement parallel processing, where a number of computing tasks are completed at the same time (in parallel).

FIGURE 10.5 Pentium Processor on a Motherboard

One method for implementing parallel processing is **pipelining**. When a task can be separated into a sequential series of subtasks to accomplish, the subtasks can be executed in parallel by different processing units. A common example of the use of pipelining is the fetch-decode-execute cycle for instruction execution. At a basic level, there are five stages:

1. Fetch the instruction (processor P1)
2. Decode the instruction (processor P2)
3. Fetch operands (processor P3)
4. Execute the instruction (processor P4)
5. Store the result (processor P5)

Five processors can each be executing one of the steps. Note that there is a defined ordering between the steps. Step 2 cannot be processed at the same time as step 1, as the instruction must be fetched before it is decoded. However, multiple instructions can be executed in a pipeline as shown in Figure 10.6.

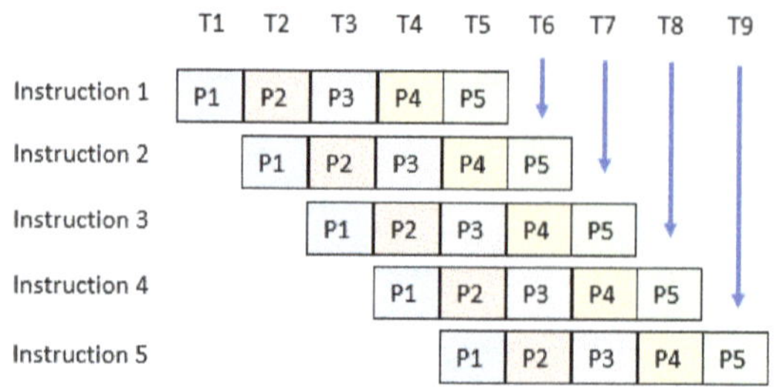

FIGURE 10.6 Five-stage Pipeline for the Fetch-Decode-Execute Cycle

Example 10.2: Five-stage fetch-decode-execute instruction pipeline in Figure 10.6.

T1: Instruction 1 enters the pipeline and is fetched.
T2: Instruction 2 enters the pipeline and is fetched.
Instruction 1 is decoded.
T3: Instruction 3 enters the pipeline and is fetched.
Instruction 2 is decoded.
Instruction 1 has operands fetched.
T4: Instruction 4 enters the pipeline and is fetched.
Instruction 3 is decoded.
Instruction 2 has operands fetched.
Instruction 1 executes.
T5: Instruction 5 enters the pipeline and is fetched.
Instruction 4 is decoded.
Instruction 3 has operands fetched.
Instruction 2 executes.
Instruction 1 stores result (now complete and is removed from the pipeline).

...

T8: Instruction 4 stores result (now complete and is removed from the pipeline).
Instruction 5 executes.
T9: Instruction 5 stores result (now complete and removed from the pipeline).

All five instructions have completed in nine time units.

In general, the time required to complete n instructions without pipelining would be $t \times s \times n$, where t is the length of each time unit, s is the number of stages in the pipeline, and n is the number of instructions. With maximum use of the pipeline, the time required is $t \times s + (n - 1) \times t$. This makes sense, as the first instruction takes $t \times s$ time units and the other $n - 1$ instructions will each take an extra time unit to complete.

Example 10.3: Computation of the amount of time needed to complete instructions in non-pipeline and pipelined systems.

- Given six stages in each instruction execution, 15 ns per time unit, and 20 instructions, compute the length of time to complete all instructions with no pipelining.

$$15 \text{ ns} \times 6 \times 20 = 1800 \text{ ns}$$

- Given six stages in each instruction execution, 15 ns per time unit, and 20 instructions, compute the length of time to complete all instructions with maximum pipelining.

$$15 \text{ ns} \times 6 + (20 - 1) \times 6 = 204 \text{ ns}$$

Note that the above is a best-case scenario for time savings. Each time unit must be the same length, which is not feasible in practice, since an operand fetch from memory will take longer than ALU execution. The time unit must represent the largest of the stages. There are also instructions that may not require all stages, such as an instruction that loads a value from memory into a register. It would not need the ALU execution, so it must do nothing and wait at that time. There might be a **data dependency** problem if, for example, instruction 2 needs to use the result of instruction 1 as an operand. At the time instruction 2's operand is fetched, instruction 1 would not have yet stored its result. Finally, there might be two steps that both need to access memory at the same time, such as an instruction fetch and an operand fetch. With a single path to memory, this will also cause a problem. This is a **resource conflict**. In these last two examples, the pipeline may need to be stalled to wait for either results or resources that are needed.

In more general circumstances, multiple processors can be used to process data at the same time to speed execution. The most common paradigms for multiprocessors to do this execution are SIMD and MIMD. **Single instruction/multiple data (SIMD)** processing represents situations where a particular task is applied to many different data items. For example, to apply the Pythagorean theorem to compute the hypotenuse for 100 input values in an array, we could use 100 processors to compute the results in parallel. Note that in this case, the processors are synchronized to do the same computation at the same time. **Multiple instruction/multiple data (MIMD)** processing is asynchronous. Multiple different tasks are scheduled on different processors. The data used for MIMD processing can be shared or distributed among the processors. In the case of MIMD, the processors must communicate with each other through message passing to efficiently schedule tasks and assemble results.

Many processors today are **multicore processors,** which utilize parallel execution. A core refers to a processing unit (CPU). A multicore processor will have at least two cores (dual core), and four cores (quad core) processors are common. Multicore processors each run their own fetch-decode-execute cycles to execute program instructions in parallel. The cores reside on the same chip, which speeds processing and shares main memory. The cores usually have their own cache and share a larger cache in a hierarchy. Figure 10.7 illustrates this structure for a dual-core processor. **Multithreading** is the separation of either an individual program or multiple programs into processes (threads) that can be executed in parallel. Each core is assigned threads to execute to implement parallelism. A multicore processor is a good example of an MIMD computer, since instructions that the cores are executing will be different.

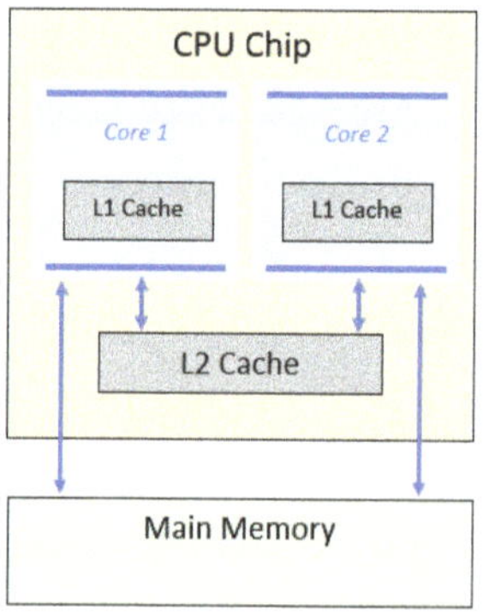

FIGURE 10.7 Dual-Core Processor

A **graphics processing unit (GPU)** is a specialized electronic circuit designed to manipulate and alter memory to accelerate the creation of images in a frame buffer intended for output to a display device. GPUs are used in embedded systems, mobile phones, personal computers, workstations, and game consoles.

In graphic computing applications, such as video games, virtual reality, augmented reality, and computer simulations, the pixels on the computer screen need to be rendered to form vivid visual objects or scenes presented to the users. Nowadays, the HD (high-definition) standard size of images, pictures, or video frames is 1280 × 720, which is close to 1 million pixels (921,600). On the hardware side, the native resolution for a basic monitor is 1080P, which is 1920 × 1080 (>2 million pixels). The resolution of 4K monitors/screens reaches 3,840 × 2,160, which have 8,294,400 pixels. Each pixel contains three values: R(Red), G(Green), B(Blue). To render an image on the screen, the total computational data is triple that of the number of pixels and involves huge computational cost. And this is just for a single image. If the frame frequency is 20 FPS, there are 20 images needing to be processed in a second. Without high performance computational power, such graphic applications are not possible.

A CPU in a computer is responsible for regular operations. If the CPU needs to handle intensive graphic computing, plus the tasks from other regular applications, the overall workload would exceed the CPU's capacity, and the computer system would work at an unacceptably slow pace. Thus, a special programmable processing unit was designed to work independently from the CPU and be responsible for graphics manipulation and output. Since this device was originally designed to be dedicated to graphic computing only, Sony used the term GPU (graphics processing unit) in reference to the PlayStation consoles in 1994. The term was popularized by NVIDIA in 1999, which is a major manufacturer of GPUs.

A GPU's parallel structure makes it more efficient than general-purpose CPUs for algorithms that process large blocks of data. Even though multicore CPUs are able to perform tasks in parallel, the difference in architecture design principle between CPU and GPU is significant. Specifically, a GPU has many more cores to process a task. Figure 10.8 shows the contrast of cores in a CPU and GPU,

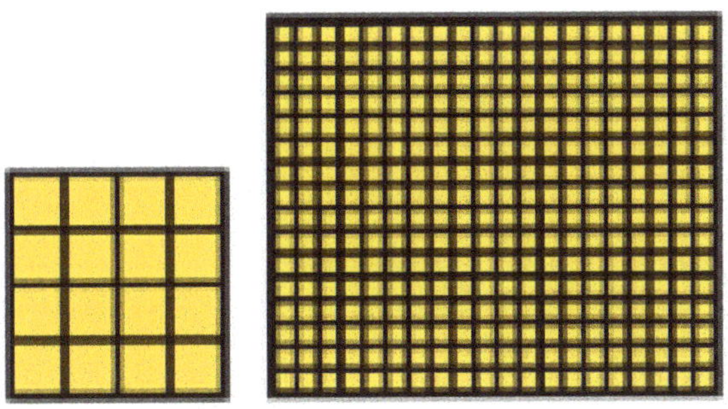

FIGURE 10.8 Cores in CPU (left) Versus Cores in GPU (right); Each Cell Represents a Core

respectively. Each core in a GPU includes an ALU (arithmetic logic unit), which is the execution component of the processor. The data processing in a GPU—many processors performing the same operation on many data items—makes it a good example of a **SIMD** system.

A common CPU is optimized to be as fast as possible to finish a task with low **latency** (delay in response). Each core in a CPU runs at a high-frequency clock speed, while a GPU is optimized for parallel data **throughput** (volume of results). Each core in a GPU runs at a clock speed significantly slower than a CPU's clock but focuses on execution throughput of massively parallel programs. That is, within a time period, many more tasks can be finished than with CPU. Figure 10.9 shows the architecture of a GPU.

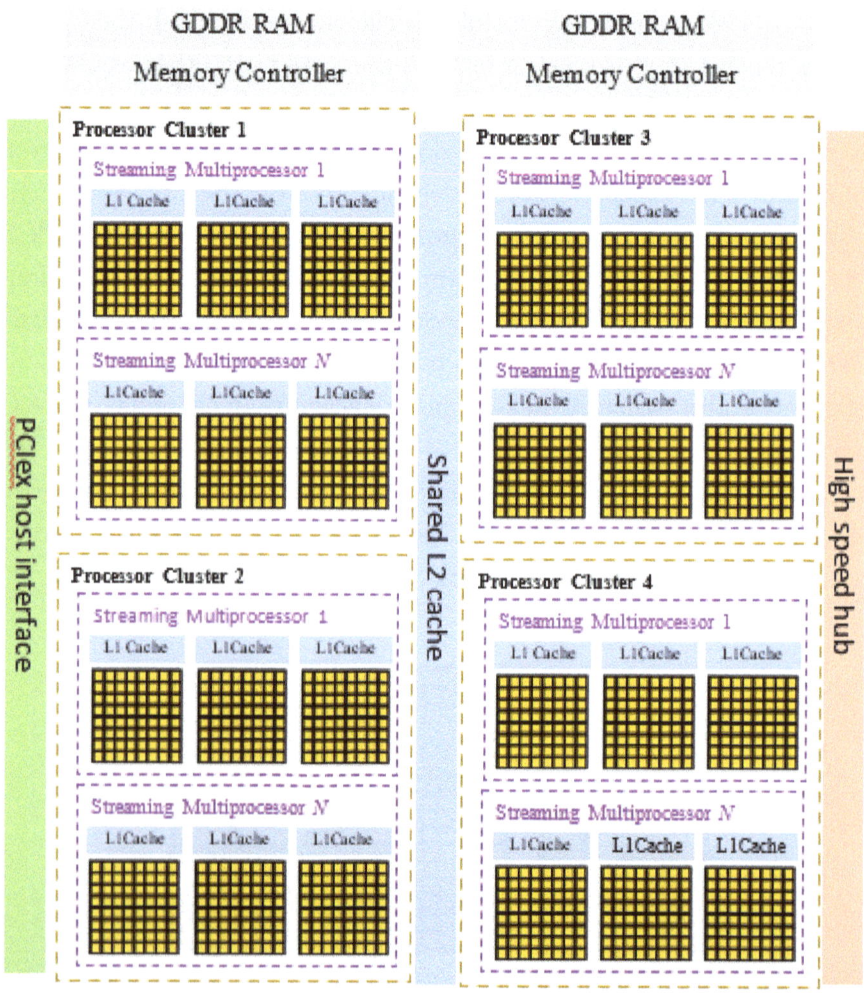

FIGURE 10.9 Example of a GPU Architecture

Compared with the CPU architecture, a GPU also uses memory cache layers, memory controller, and global memory, but **with relatively fewer and small memory cache layers.** In the diagram, a single GPU device consists of multiple processor clusters (PCs) that contain multiple streaming multiprocessors (SM). Each yellow grid cell represents a single core. Each SM accommodates a layer-1 cache with its associated cores. Typically, one SM uses a dedicated L1 cache and a shared layer-2 cache before pulling data from global GDDR memory. Its architecture is tolerant of memory latency due to the cache structure. Using NVIDIA Tesla V100 GPU as an example, it contains 80 streaming multiprocessors (SM), each containing 64 cores, making a total of 5,120 cores! Tasks for GPUs are not scheduled to individual cores, but to processor clusters and SMs. That's how it's able to process in parallel.

GPUs were originally designed to accelerate the rendering of 3-D graphics. It allowed graphics programmers to create interesting visual effects and realistic scenes with advanced lighting and shadowing techniques. GPUs were best known for their capabilities in gaming. Later, other developers also began to tap the power of GPUs to dramatically accelerate additional workloads in high-performance computing (HPC), deep learning, and more.

Now, modern GPUs are not only efficient at manipulating computer graphics and image processing, but also more efficient than general-purpose central processing units (CPUs) for algorithms that process large blocks of data in parallel. GPU has become one of the most important types of computing technology, both for personal and business computing. It is becoming more popular for use in creative production and artificial intelligence (AI).

Chapter Summary

The CPU has the job of executing programs through use of the fetch-decode-execute cycle. Binary machine language instructions specify an operation, given by an opcode, to be performed and references to operands needed by the instruction. The PC register stores the address of the next instruction to be executed. The instruction at that address is loaded into the IR and decoded. The PC is incremented at this time to address the next instruction. The decode step determines the operation to perform based on the opcode in the instruction. In the execute phase, operands are fetched, the operation is applied to the operands, and the result is stored. The CPU uses a memory address register, MAR, and a memory buffer register, MBR, to communicate with memory. These registers are needed for fetching instructions and operands from memory and storing result values to memory. If a branch instruction has been processed, the PC is updated to the new program address to be executed. The process repeats by fetching the next instruction (based on the updated PC) and continuing.

The control unit guides the execution of the fetch-decode-execute cycle. This includes the process or decoding instruction, fetching instructions and operands from memory, driving the ALU, and communicating with memory. The control unit may be hardwired or microprogrammed. A hardwired control unit utilizes digital logic for control of computer operation. A microprogrammed control unit uses microcode, a simple language for performing basic steps within a machine language instruction. Microprogrammed control uses a microprogram counter (MPC) similar to a PC to control its execution.

The ALU performs arithmetic and logic operations needed by the CPU. The digital logic in the ALU starts processing by using a decoder that will choose which operation is executed. Typical operations performed by an ALU include arithmetic (incrementing, addition, subtraction, multiplication, division), logical (AND, OR, NOT, EX-OR), and bit shifting and rotation operations. The digital logic for an ALU will process all operations and use a multiplexer to select the output from the desired operation. It is common to build logic for the ALU, which will process a single-bit operation and repeat (ripple) the ALU units to perform processing on many bits.

Increasing the number of processors in a computer system can significantly speed processing by executing tasks in parallel. Pipelining is one method for introducing parallelism into a computer system. In pipelining, a task is separated into a set of steps that must be done in sequence. If there are multiple instances of the task to be completed, each instance can be started in the pipeline in turn. One step from each instance can be executed at the same time. A pipeline can be made of instructions in a program using the steps in the fetch-decode-execute cycle as pipeline stages. Two common paradigms for parallel execution are SIMD and MIMD. SIMD systems synchronize multiple processors to execute the same instruction on many data items in parallel. A GPU is an example of a SIMD machine. MIMD systems execute different instructions in an asynchronous fashion in parallel. Message passing is used to efficiently distribute instructions to the processes and organize the results of the computations. Multicore processors are MIMD systems with multiple cores on a chip. Each core runs its own fetch-decode-execute cycle on threads that it is scheduled to process.

Chapter 10 Exercises

1. The PC and IR registers are related to one and other but serve different purposes. Describe how they are related and how they differ.

2. The MAR and MBR registers are related to one and other but serve different purposes. Describe how they are related and how they differ.

3. The PSR has a value of 1 for the overflow bit. Describe how this may have been caused.

4. Assume that the opcode portion of an instruction is made up of 3 bits.

   a. How many different instructions can this represent?
   b. Given your answer to 4a, list a set of that many possible instructions.

5. Give an advantage of hardwired control over microprogrammed control.

6. Give an advantage of microprogrammed control over hardwired control.

7. Consider the 1-bit ALU show in Figure 10.3. Fill in the blank portions of the following truth table. If there are several possible values in a cell, include both.

F1	F0	A	B	OUTPUT
0	0		0	0
1	0	0	1	
1	1		0	0
0	1	1		1
1	1	1	1	
0	1	0		1
1	0		1	1
0	1	1	0	
1	0		1	1
1		1	0	0

8. Consider the four AND gates at the left of Figure 10.3.

   a. Show the two inputs and outputs from each of these gates given F1 = 1, F0 = 0.
   b. Explain in detail how these four outputs will cause the output of A X-OR B to be the result of the ALU.

9. Assume that there are five distinct stages in the execution of an instruction. Each stage takes 12 ns to execute.

   a. Compute how long it would take to execute 10 instructions with no pipelining.
   b. Assume that there are no data dependency or resource conflict problems. Compute how long it would take to execute the 10 instructions with pipelining.
   c. Show sequence of three assignment statements of the form ***var1*** = ***var2*** + ***var3*** which represent a good example of a data dependency in a pipelined system.

10. Amazon just received a million orders from New York state customers and needs to compute the tax for each. Would this be a problem more appropriate for a SIMD or MIMD multiple processor system. Explain your answer.

11. A quad core processor is an example of a MIMD system. Explain why such a processor fits nicely into this category.

Credit

Fig. 10.5: Capitalistcakes, "Pentium processor on a motherboard," https://commons.wikimedia.org/wiki/File:Intel_Pentium_Pro_256k.jpg, 2019.

CHAPTER

11

I/O Processing

Introduction

Input/Output (I/O) processing allows the CPU to communicate with I/O devices, including memory, to read and write data. A **bus** is used to transfer data between the CPU and other devices, with conduits for moving control information, addresses, and data.

The control unit in the CPU determines what data needs to be moved where and directs the traffic along the bus. A **handshaking** process (control information passed between the CPU and the I/O device) is used to request, reserve, and release the device. There are decisions to be made about where data is stored during movement and when the data is moved along the bus for maximum efficiency and reliability. A **system clock** can be used to ensure that a bus transfer has been completed before another transfer begins. **Interrupts** (signals generated by the CPU or a device) can be used to indicate when data is ready to be transferred or a transfer has been completed.

Objectives

This chapter will help students:

- **understand** the need for control, address, and data movement on a bus.
- **explain** the uses of a clock in a computer system.
- **illustrate the** I/O process of handshaking.
- **describe** programmed I/O and interrupt-driven I/O.

- **summarize** memory-mapped I/O.
- **understand** the interrupt process.

Key Terms

bus: conduit for moving data between components of a computer system

system clock: a computer component that sends out high and low pulses to facilitate timing

handshaking: I/O data transfer communication between the CPU and an I/O device

polling: repeated requests to a device to determine its status

programmed I/O: I/O transfer controlled by the CPU using polling of the device

interrupt: a signal sent by the CPU or a device that stops execution of the current program

interrupt-driven I/O: I/O that is based on interrupts indicating when data transfer is completed

memory-mapped I/O: I/O transfer where the CPU sees data on a device as an extension of memory

Bus

A bus is used to transfer data between components of a computer. Only one data item of a given number of bits determined by the bus capacity can travel on the bus at a time. The control unit in the CPU manages the traffic that moves along a bus. The system bus is the backbone of communication in a computer system, as shown in Figure 11.1. The components shown in the figure include the CPU and RAM; ROM, which contains microcode and other constant system information; I/O, which can include various input/output devices, such as a monitor, keyboard, or printer; and perhaps other peripherals. The goal of the

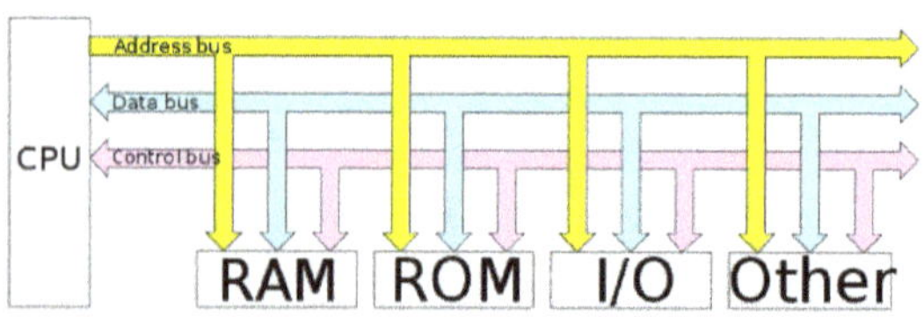

FIGURE 11.1 Computer System Bus

use of the bus is to accomplish the reading or writing of data between the CPU and another device.

The **bandwidth** of a bus defines the amount of data that can be transferred in one second. This value will be based on the **cycle rate** of the bus (number of transfers per second) and the **data width** of the bus (number of bits transferred in one cycle). The formula for computing the bandwidth is:

$$\text{bandwidth} = \text{cycle rate} \times \text{data width}$$

So, if we have a cycle rate of 100 MHz (100 million cycles per second) and a data width of 2 bytes:

$$\text{bandwidth} = 100 \text{ MHz} \times 2 \text{ bytes} = 200 \text{ Mbyte/s}$$

There are three types of information sent by the CPU, with a bus associated with each. The address bus sends the address of the data being accessed from the CPU to the device. The MAR in the CPU will be loaded with the address to be sent on the bus. The data bus will transfer the data associated with the command. The MDR in the CPU will receive (in the case of a read instruction) or send (in the case of a write instruction) the data to be moved on the bus. The control bus will transfer information, ensuring the correct operation of the bus. This information includes an indication of the type of transfer (read or write), synchronization signals, and signals, ensuring that only one transfer is occurring on the bus at that time.

There is significant data traffic necessary on the system bus, including the fetching of commands and operands from memory, writing results, and data movement to I/O devices. Many bus accesses may be necessary to execute a single machine-language instruction. A single system bus can slow the execution of programs due to the bottleneck caused by this heavy traffic. A number of multi-bus systems have been developed to alleviate the traffic on an individual bus.

Clock

A system clock is necessary to control the operation of a computer system. The clock sends out pulses at a very high rate of speed. It sends out a sequence of low-high signals that can be looked at as binary values of 1 and 0, as shown in Figure 11.2. The cycle time is the time between clock ticks.

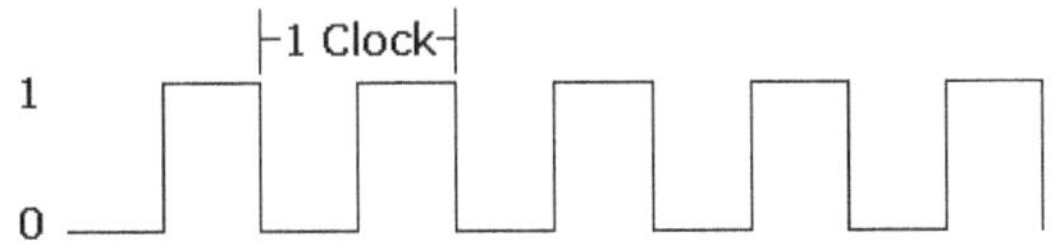

FIGURE 11.2 System Clock Low-High Signals

The speed of a clock is measured in gigahertz (GHz). A 3 GHz processor will have a clock that ticks 3,000,000,000 times per second. As a rule, the faster the cycle time, the faster the processing of instructions. The clock is used to provide timing signals for a number of processes that the computer is performing. For example, the clock is used to control the speed at which CPU instruction steps (either hardwired or microprogrammed) are executed. This ensures the completion of each step before the execution of the next. Similarly, the clock controls memory access speeds so that the memory system updates information reliably. Remember that one of the inputs to a flip-flop is CLK. CLK will have a value of 1 if the clock value is high and the address of the flip-flop is chosen. Therefore, if the CLK is low, memory will be blocked from changing, ensuring correct timing.

Interrupts

An interrupt is a signal that stops the execution of a program and transfers control to the operating system. The source of the interrupt can be generated by a currently executing program or from a connected device. Program-generated interrupts might be caused by exceptions (such as dividing by zero or a page fault) or requests that the program makes of the operating system (such as an I/O request). Since interrupts stop the execution of the current program, they are addressed immediately.

There is a portion of the operating system code that addresses interrupts, called an **interrupt handler**. When an interrupt occurs, a code that describes the cause and source of the interrupt is written to a reserved portion of memory along with the current value of the PC. The PC must be saved since the program that was interrupted will, in most instances, need to continue after the interrupt has been processed. The coded control structure of the interrupt handler is very simple. It is a set of statements checking for each possible type of interrupt. Within each if statement, there will be code that processes that type of interrupt.

I/O Processing

The I/O system allows communication of data between the computer system and external devices. Some common I/O devices include the disk drive, monitor, mouse, keyboard, and printer. The system bus, as shown in Figure 11.1, is used to transfer data to these devices. There is a **handshaking** process that initiates and ends communication between the CPU and an I/O device to coordinate the data transfer. The CPU communicates with the device to make sure that it is available. If available, the device will set a busy bit to indicate that it is now reserved for the requested data movement and signals back to the CPU that the transfer can be made. Once the request has been satisfied and the data moved, the device resets the busy bit and once again becomes available.

There are several control methods for the transfer of data between the computer and the external device.

programmed I/O uses a port (connection) for each I/O device. The port contains a control register that indicates whether data is ready to be accessed. An example of such a device might be a keyboard. The CPU periodically polls the control register to see if a character has been typed. When the control register ready bit has been set, the character can be retrieved from a register in the port. The ready bit is reset. In the case of programmed I/O, the CPU has control over the processing of I/O based on its algorithm for polling devices.

interrupt-driven I/O has the CPU send I/O requests to devices. The device processes the request. The CPU can continue working on other tasks during the time that the device is satisfying the request. The device will signal an interrupt when the request has been completed. At that time, the CPU can process the data associated with the request. This removes the need for polling the devices to determine if I/O requests have been satisfied.

programmed I/O is simpler to implement. The CPU has control over the device polling process and the order in which devices will be processed. Extra devices can be added, and polling priorities can be changed easily. In general, interrupt-driven I/O will process I/O more efficiently since devices will be handled as soon as they are ready. CPU resources will also not be kept busy polling devices.

memory-mapped I/O (a form of interrupt-driven I/O) reserves an area of memory on the computer (RAM) and an area of memory on the I/O device. The I/O device memory is seen as an extension of RAM. The I/O device memory can be accessed directly in the same way that the CPU would access RAM. With memory-mapped I/O, the CPU does not need to be directly involved with the movement of data. It sends a request to the device to transfer data. A control unit on the device will perform the movement of data. The CPU can continue working on other tasks until the request is satisfied, at which point an interrupt is signaled and the data can be accessed from memory. The advantage of using a memory-mapped I/O system is that adding new I/O devices is simplified since the protocol for communication (using RAM-like memory access) is the same for all devices.

Chapter Summary

It is necessary for a computer system to have efficient and reliable communication between the CPU and other components in the system, including computer memories and I/O devices such as a keyboard, monitor, mouse, printer, or scanner, among others.

A bus is the conduit for communication between the components in the system. The bus has a data width that gives the number of bits in a transfer. The bandwidth of the bus defines the speed at which bytes are transferred per second. A bus system must transmit not only data but also address information (where the data will be sent or retrieved from) and control information (read/write indicator and status information). A clock sends out regular low-high signals at a very high rate of speed. The clock is fused for timing and data access associated with I/O transfer and the CPU.

The operating system needs to be involved when reading and writing data, both in memory operations and in communication with I/O devices. Interrupts are used to immediately send program control to the operating system's interrupt handler to deal with I/O request processing. Common sources for interrupts associated with I/O processing include a device being made available and data being ready for transfer.

When a program wants to move data to or from a device, a handshaking process is initiated. This communication between the CPU and the device assures that the device is available for use. There are two main methods for coordinating the transfer of data. Programmed I/O gives the CPU control over when data is transferred. The CPU periodically polls devices to see if they are free for transfer or have data available for movement. This process gives the CPU control over when data is moved but uses extra CPU resources and may not move data at the time that it is ready for transfer. Interrupt-driven I/O has devices that use interrupts to indicate when a device or data is ready. Programmed I/O has the advantages of simplicity and CPU control over device data transfers. Interrupt-driven I/O has the advantage of efficiency of processing device requests and does not need to spend CPU resources polling devices.

Memory-mapped I/O allows the CPU to view I/O device data as an extension of main memory. This allows new devices to be easily added to the system, as CPU reference to their data will be made in the same way as other devices.

Chapter 11 Exercises

1. Explain why buses need to carry more than just data to devices.
2. Explain the effect of bandwidth on the efficiency of a computer system.
3. Describe the two factors in computing the bandwidth of data movement in a computer system.
4. Explain how interrupt-driven I/O gives a quicker response to I/O requests than programmed I/O.
5. Describe the two basic steps that the operating system performs when an interrupt occurs.
6. Describe memory-mapped I/O.
7. Explain the similarity between the use of memory-mapped I/O for two apparently very different devices, such as the keyboard (an input device) and the monitor (an output device).

CHAPTER

12

Instruction Set Architecture

Introduction

This chapter covers the basic concepts of instruction set architecture (ISA). Two major types of ISA, reduced instruction set computers (RISC) and complex instruction set computer (CISC), are introduced.

Objectives

This chapter will help students:

- **understand** the concepts of computer instruction set architecture.
- **illustrate** the current types of ISA.
- **identify** the difference between the two major ISAs (RISC and CISC).

Key Terms

ISA: instruction set architecture

little endian: the least significant byte of the data is placed at the byte with the lowest address

big endian: the most significant byte of the data is placed at the byte with the lowest address

CISC: complex instruction set computer

RISC: reduced instruction set computer

Introduction to Instruction Set Architecture

An ISA as part of the abstract model of a computer defines the computer instructions and the format of each instruction. The ISA acts as an interface between the hardware and the software. The defined group of commands and operations is used by the software to communicate with the hardware. Software is converted to machine instructions using a compiler/interpreter. Then the machine instructions are executed directly on the hardware.

An ISA contains the functional definition of storage locations (registers, memory), operations (add, multiply, branch, load, store, etc.), and precise descriptions of how to invoke and access them. An ISA does not contain nonfunctional aspects, such as how operations are implemented, which operations are fast, which operations are slow, which operations take more power, and which operations take less power.

A computer architect has many decisions to make when encoding an instruction set, such as choosing the operations in the instruction set, the number of bits per instruction (16, 32, and 64 are the most common), the number of operands per instruction, and the instruction/data type.

The ISA of a processor has the following five categories:

- **operand storage in the CPU:**
  Where are the operands stored other than in memory? They can be stored in a stack structure or in registers.
- **number of explicit named operands per instruction:**
  How many operands are named in a typical instruction? Zero, one, two, and three are the most common numbers of operands.
- **operand location:**
  Can any ALU instruction operand be in memory, or must all operands be in the CPU? Instructions can be classified into three types: register-to-register, register-to-memory, and memory-to-memory. The combination of different operands allowed in each instruction must be defined.

- **operations:**
  Which operations are provided in the ISA? This includes different operation types and the instructions that can/cannot access memory.
- **type and size of operands:**
  What is the type and size of each operand, and how is it specified? Operands can be addresses, numbers, or even characters.

There are four main factors used to measure ISAs, including: (1) the amount of space required by a program; (2) the complexity of the instruction set and the complexity of the tasks performed by the instructions; (3) the length of instructions; and (4) the total number of instructions. The following key points are conducive to designing an instruction set:

- Short instructions occupy less memory space and can be fetched quickly. Although the speed of execution of short instructions is better than that of long instructions, they limit the power of each instruction and the size and number of operands. Short instructions cannot always meet the needs of a robust design of an instruction set.
- Although fixed-length instructions are easier to decode, they waste space.
- The instruction format is affected by memory organization. It is difficult to access any single character of the stored word of different lengths in non-byte-addressable memory. Therefore, machines with different-size words are often byte-addressable to ensure every byte has a unique address.
- A fixed-length instruction does not mean a fixed number of operands. Different numbers of bits in the operand field are allowed in an ISA with a fixed overall instruction length.

Little Endian Versus Big Endian

In computer architecture, the term *endian* refers to "byte order." How are the bytes of a multiple-byte data element stored in computers?

> little-endian byte order: The least significant byte (LSB, the "little end") of the data is placed at the byte with the lowest address.

> big-endian byte order: The most significant byte (MSB, the "big end") of the data is placed at the byte with the lowest address.

Most PCs are little-endian machines. Most UNIX machines are big-endian machines, as shown here:

Byte 3	Byte 2	Byte 1	Byte 0

Example 12.1: Consider an integer requiring 4 bytes.

On a little-endian machine, this is arranged in memory as follows:

Base Address + 0 = Byte0

Base Address + 1 = Byte1

Base Address + 2 = Byte2

Base Address + 3 = Byte3

On a big-endian machine, this long integer would then be stored as:

Base Address + 0 = Byte3

Base Address + 1 = Byte2

Base Address + 2 = Byte1

Base Address + 3 = Byte0

For instance, the decimal number 6,626,874 is converted to the hexadecimal number 651E3A. Each digit requires a nibble, so 1 byte holds two digits. As shown in Figure 12.1, this hex value 651E3A is stored in memory in little- and big-endian byte order, respectively.

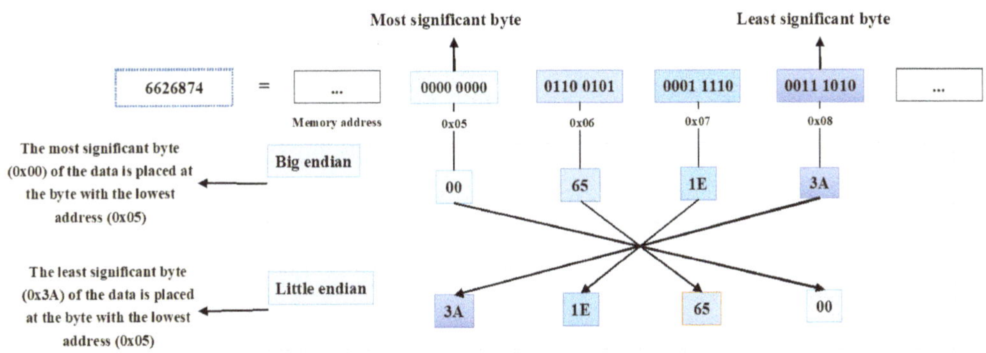

FIGURE 12.1 Big-Endian and Little-Endian Expressions of the Decimal Number 6,626,874

Big-endian has the following four main advantages:

1. Reading the arithmetic sign bit of the wrong word size, integer size, or floating-point number size always yields a correct value.

2. Most network header code and bitmap graphics (display and memory arrangement) are mapped using the "left MSB" scheme, which means that the shifting and storage of graphic elements larger than 1 byte are handled naturally by the architecture. As a major performance disadvantage of little-endian machines, reversing the byte order is required when processing large graphics elements.
3. When decoding variable-length bit codes (compressed data), such as Huffman coding and the Lempel–Ziv–Welch (LZW) algorithm, if the MSB is encoded as LSB (big-endian), the code word is used as an index of a lookup table. The same goes for encoding, because the shifted bits must be mirrored to generate such codes on a little-endian machine.
4. It is easy to read hex dumps.

Big-endian has the following two main disadvantages:

a. Reading a value with the wrong word length results in a wrong value. When done on a little-endian architecture, it can sometimes yield a correct result.
b. Most big-endian architectures (non-Intel) do not allow writing words on non-word address boundaries (odd addresses). Intel allows odd address reads and writes (they are split into two separate operations), which makes it easier for programmers but more difficult for hardware designers.

The pros and cons of little endian are basically the opposites of those listed above.

Any program that writes data to or reads data from a file must know the byte order on a particular machine. For instance, the Windows BMP graphics format was developed on little-endian machines. When a BMP is reviewed by an application on a big-endian machine, the corresponding byte order needs first to be reversed. The byte-ordering issues are considered by software designers. For example, Adobe Photoshop, JPEG, MacPaint, and Sun Raster files use big endian. GIF, PC Paintbrush, and RTF by Microsoft use little endian. Microsoft WAV and AVI files, TIFF files, and XWD (X Windows Dump) support both big endian and little endian, typically by encoding an identifier into the corresponding file.

Internal Storage in the CPU

Instructions are in bit patterns. Hardware interprets these as commands. Hardware determines the instruction format. Internal storage space in a process is a factor that affects an ISA. Instructions used to execute operations take one or more operands.

The ISA can be classified as stack architecture, accumulator architecture, or register-set architecture, based on where the operands are stored and whether they are named explicitly or implicitly.

Stack

The **stack** architecture, as shown in Figure 12.2, was used in computers in the 1960s. The operands are implicitly put into the stack. Operations take place at the top two locations on the stack, including removing the operands and leaving the result on the top of the stack. Although the stack is a block of memory in RAM, the CPU has a stack pointer register that addresses the top of the stack. The first-in and last-out strategy is accomplished using two operations, push and pop, and each has one operand. Push adds its operand value to the top of the stack, and pop removes the top element of the stack and writes to its destination. Other operations involving the top element(s) of the stack have implicit operands. To perform an operation, the operands are pushed onto the stack first. Once an operation is performed, its operand(s) is deleted from the stack, and its result is stored at the top of the stack. Then, the pop command is applied to return the result back to the memory.

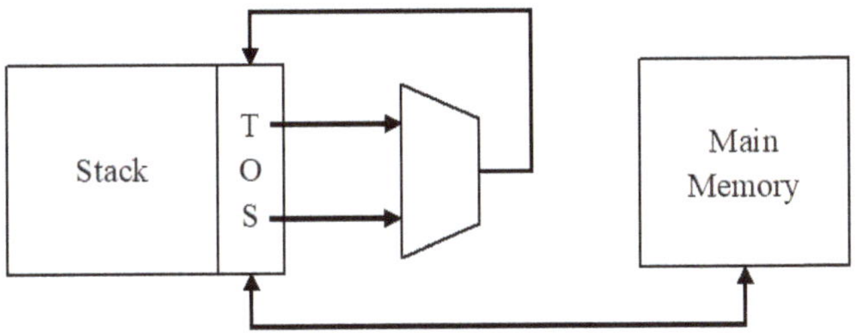

FIGURE 12.2 An Illustration of a Stack. TOS Stands for "Top of Stack."

Example 12.2

Suppose a stack architecture machine executes the operation C = A + B, where A, B, and C are memory addresses. The result of A + B is stored at location C. The following steps are used to implement the operation C = A + B in a stack architecture:

1. **PUSH A** # push value at location A to the top of the stack
2. **PUSH B** # push value at location B to the top of the stack
3. **ADD** # remove and add the top two elements of the stack, and save the result back to the top of the stack
4. **POP C** # store the value at the top of the stack (which is the result) to location C

Accumulator

The **accumulator** (ACC) architecture, as shown in Figure 12.3, was used in early computers like the IBM 7090. Now it is only used by some digital signal processors. One operand used by an instruction implicitly comes from the accumulator. The other operand is explicitly specified by an address in memory. The result of ALU is written back to the accumulator. The command LOAD is used to load the operand in the accumulator from memory. The command STORE is used to store the result from the accumulator in memory.

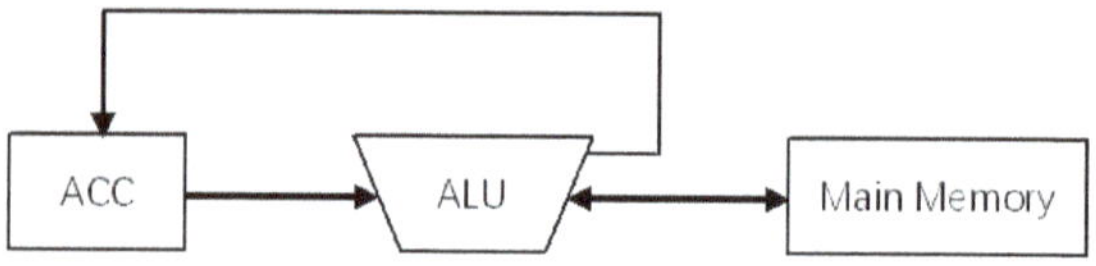

FIGURE 12.3 Illustration of an Accumulator

Example 12.3

Suppose an accumulator architecture machine executes the same operation C = A + B as above. The following steps are used to implement the operation C = A + B in an accumulator architecture:

1. **LOAD A** # load value at location A from main memory to ACC (accumulator)
2. **ADD B** # fetch the value at location B, add it to the value at ACC (which is the value A), and save the result back to ACC
3. **STORE C** # store the value at ACC (which is the result of A + B) to location C

Register-set Architecture

The **register-set architecture** is the dominant architecture used by modern computers. A number of general-purpose registers (GPRs) are used for internal storage in modern CPUs. The number of GPRs in a CPU generally ranges between 8 and 32. This architecture allows fast access to temporary values and permits clever compiler optimization. But the related instructions are longer than the ones used in the accumulator designs. All operands used in operations are explicitly specified. The operands are available in either memory or registers, which can be classified as register-register architecture, register-memory architecture, or memory-memory architecture.

In the *register–register* architecture, as shown in Figure 12.4, the registers are used for storing operands. Since only load and store instructions have memory operands, this architecture is also called load-store architecture. Most RISCs adopt this architecture.

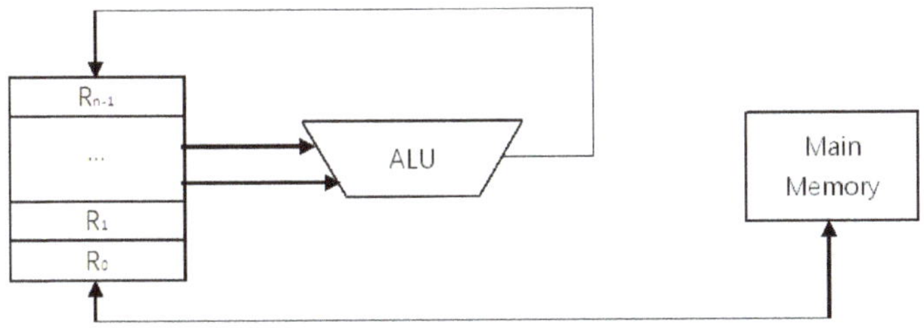

FIGURE 12.4 Illustration of Register-Register Architecture

RISC and CISC

RISC and CISC are two different designs of ISA. RISC is an acronym for reduced instruction set computer, and CISC is short for CISC. The ISA specifies communication protocol between the hardware processor and the software programs. The earliest computers had a CISC architecture. Later in the 1980s, RISC architecture was gradually developed to overcome the increasing complexity of CISC processors.

RISC

A RISC microprocessor architecture utilizes a small, highly optimized set of instructions rather than the highly specialized set of instructions typically used in other architectures. As an alternative to the CISC architecture, RISC is often considered the most efficient CPU architecture today.

With RISC, simplified instructions do less but can execute more rapidly. The corresponding processor design principle of these simplified instructions can be implemented by a central processing unit (CPU), thereby improving performance. As a key feature, RISC allows developers to increase both the register set and internal parallelism by increasing the number of parallel threads executed by the CPU and the speed of the CPU's executing instructions. Advanced RISC machines (ARM) are developed based on RISC. ARM architecture-based processors are widely used in smartphones, tablets, laptops, gaming consoles, and other smart devices.

Considering energy efficiency, RISC can provide high performance per watt, especially for battery-operated devices. One action per instruction is executed in a RISC processor. The optimization of execution time can be done in just one cycle. According to the fixed length of instructions, RISC architecture is easy to pipeline. Due to the lack of complex instruction decoding logic, RISC supports more registers and spends less time on loading and storing values to memory. Because of the reduced instruction set and simple decoding logic, chip designers can use less chip space, fewer transistors, and more GPRs to simplify CPU design and implementation.

CISC

A CISC is a computer where single instructions can execute several low-level operations (such as a load from memory, an arithmetic operation, and a memory store) and use multiple addressing modes.

CISC ISAs use variable-length instructions. Therefore, simple instructions are short, similar to RISC, but more complicated instructions that have a longer length are also allowed. In CISC architecture, a large number of instructions can directly access memory. For the variable-length instructions, the number of clock cycles per instruction varies. Hundreds of cycles may be required for some complex instructions. The execution time is extended when the system clock slows down to allow sufficient time for instructions to complete in certain circumstances.

Comparison of CISC and RISC

As shown in Table 12.1, compared with RISC, CISC is highly efficient in terms of the number of instructions per program, thereby resulting in lower demand for code storage and RAM. Additionally, a CISC processor contains more transistors to process complex instructions.

TABLE 12.1 Comparison of CISC and RISC

CISC	RISC
Emphasis on hardware.	Emphasis on software.
Includes multi-clock complex instructions.	Includes single-clock and only reduced instructions.
Uses a memory-to-memory method. "LOAD" and "STORE" are incorporated in instructions.	Uses a register-to-register method. "LOAD" and "STORE" are independent instructions.
Involves small code sizes and high cycles per second.	Involves large code sizes and low cycles per second.
More transistors are used for storing and processing complex instructions.	More transistors are used in CPU registers.

The following equation is used to measure computer performance, which is directly proportional to clock cycle time, the number of clock cycles per instruction, and the number of instructions in the program.

$$\text{Time/program} = \text{time/cycle} \times \text{cycles/instruction} \times \text{instructions/program}$$

The CISC machines attempt to minimize the number of instructions per program, sacrificing the number of cycles per instruction. The RISC machines do the opposite, reducing the cycles per instruction at the cost of the number of instructions per program.

CISC machines use microcode to process each instruction, thereby managing instruction complexity. Microcode is compact and efficient, but its efficiency is affected by variable-length instructions that slow down the decoding process. Additionally, a varying number of clock cycles per instruction also introduces difficulties in pipeline implementation of instructions. Moreover, each instruction interpreted by microcode is fetched from memory and needs extra time for an additional translation process. The complex instruction set takes additional time to look up instructions and engage suitable hardware for execution.

In contrast, most RISC instructions execute in one clock cycle. So, hardwired control is used to replace microprogrammed control to achieve fast instruction execution. Due to the simplicity of the instructions, RISC machines require fewer transistors of hardware space than the complex instructions, leaving more room for GPRs. As a "side effect" of the simplicity, more instructions are required for a same operation; thus, RISC's program size (code size) is usually larger than CISC's, which could increase the workload for the RISC code compiler. During the execution stage, to increase the efficiency, instruction pipelining is a possible solution, as all of the instructions execute in a uniform amount of time (i.e., one clock).

Example 12.4

To compute the product, 10 × 15, the code on a CISC machine is given as follows:

```
mov ax, 15
mov bx, 10
mul bx, ax
```

Since there is no multiplication instruction in a minimalistic RISC ISA, the code on a RISC system is given as follows:

```
        mov ax, 0
        mov bx, 15
        mov cx, 10
Begin:  add ax, bx
        loop Begin        #causes a loop cx times
```

The CISC code is shorter but requires more clock cycles to execute. Suppose each register-to-register move and each loop operation consumes one clock cycle, respectively, on both architectures. Additionally, suppose a multiplication operation requires 30 clock cycles. The corresponding total time cycles of CISC and RISC are obtained as follows:

CISC instructions:

Total clock cycles = (2 **movs** × 1 clock cycle) + (1 **mul** × 30 clock cycles)
= 32 clock cycles

RISC instructions:

Total clock cycles = (3 **movs** × 1 clock cycle) + (10 **adds** × 1 clock cycle) + (10 **loops** × 1 clock cycle)

= 23 clock cycles

RISC clock cycles are usually shorter than CISC clock cycles. In this case, even though there are more instructions, the actual execution time for RISC is less than that for CISC.

Chapter Summary

This chapter discusses ISA, its five categories, and the four main factors used to measure ISAs. Little-endian and big-endian byte orders are explained and illustrated. The pros and cons of each byte order are discussed. Internal storage architectures, stack architecture, accumulator architecture, and register-set architecture are discussed as well. RISC and CISC are explained and compared.

Students should understand ISA, little-endian and big-endian byte orders, stack architecture, accumulator architecture, register-set architecture, RISC, and CISC. They should also be able to apply the related knowledge to understand different computer architectures.

Chapter 12 Exercises

1. Let's assume there is a byte-addressable 32-bit machine. Now show the memory space arrangements (orders) for the following hexadecimal words (4 bytes) by using little-endian and big-endian formats, respectively.
   Note: suppose the memory address starts with 0xFF00

   a. 0x1234ABCD

	address 0xFF00	address 0xFF01	address 0xFF02	address 0xFF03
Big Endian				
Little Endian				

   b. 0x11EE22FF

	address 0xFF00	address 0xFF01	address 0xFF02	address 0xFF03
Big Endian				
Little Endian				

   c. 0x00115577

	address 0xFF00	address 0xFF01	address 0xFF02	address 0xFF03
Big Endian				
Little Endian				

   d. 0x12345678

	address 0xFF00	address 0xFF01	address 0xFF02	address 0xFF03
Big Endian				
Little Endian				

2. What kinds of problems will be caused if you transfer data from a big-endian machine to a little-endian machine?

3. Show a sequence of assembly language instructions that will compute the following (using the correct precedence of operations) for the given architectures:

$$F = A + B * C - D / E$$

   a. Using a stack architecture
   b. Using an accumulator architecture
   c. Using a register-set architecture

4. Compute the Time/Program measure of computer performance for the following cases:

   a. time/cycle = 4 ns cycles/instruction = 3 instructions/program = 300
   b. time/cycle = 8 ns cycles/instruction = 5 instructions/program = 85

5. Given question 4, which example (a or b) would represent a CISC machine? Which example would represent a RISC machine? Explain your answer.

CHAPTER

13

A Simple 16-bit RISC Instruction Set for a CPU Simulator

Introduction

As mentioned in the previous chapter, a CPU has a set of embedded instructions provided by its manufacturer. In this chapter, we introduce a CPU simulator, the SimX system, and a set of instructions to illustrate the low-level computer operations. Additionally, the corresponding assembly language and programming examples are explained. After studying this chapter, students will have a better understanding of the working mechanism of a computer system and preliminary knowledge of assembly programming.

Objectives

This chapter will help students:

- **understand** a basic simulator for a simple instruction set.
- **convert** binary encodings into instruction mnemonics.
- **understand** an assembly program for arithmetic calculation.
- **identify** different classes of instructions: zero-, single-, and two-operand instructions.

Key Terms

assembly program: a set of instruction mnemonics to organize the coding statements. This type of program has direct mapping with the binary encodings

assembler: a program that can compile and translate an assembly source program into assembled object files in binary

linker: a software that takes multiple assembled object files and merges them into a single object file or executable binary code

Von Neumann Architecture

According to the Von Neumann architecture, a computer system consists of CPU, and I/O devices connected by a bus system. The programs (software) are initially stored in the secondary storage device, e.g., hard-drive disks. When a program needs to be executed, it is first loaded into main memory. Then, the CPU can run it by performing execution cycles until the program is finished. Each execution cycle has three steps:

1. The CPU first **fetches** the data or instruction from the main memory. The size of fetched information is one word. The word size is dependent on the computer hardware system. Currently, the 64-bit CPU is the mainstream hardware system. Each word has 64 bits or 8 bytes. Previously, 32-bit systems were predominant with 32-bit (4-byte) words. Similarly, in a 16-bit system, one word has only 2 bytes. In this chapter, for the sake of simplicity, a 16-bit system is used for simulation purposes.
2. Once the data is fetched by the CPU, the decoder of the control unit is triggered to perform decoding. It converts the fetched binary data into an instruction. As explained in the previous chapter, instructions are the low-level operations that are encoded by the CPU manufacturer. CPU is able to understand and perform those instructions. This decoding step can be treated as the translation procedure.
3. Finally, the translated instruction is executed by the CPU, and results of the execution may be stored into a register or memory space, depending on the definition of the instruction.

Figure 13.1 shows the overall procedure of program execution.

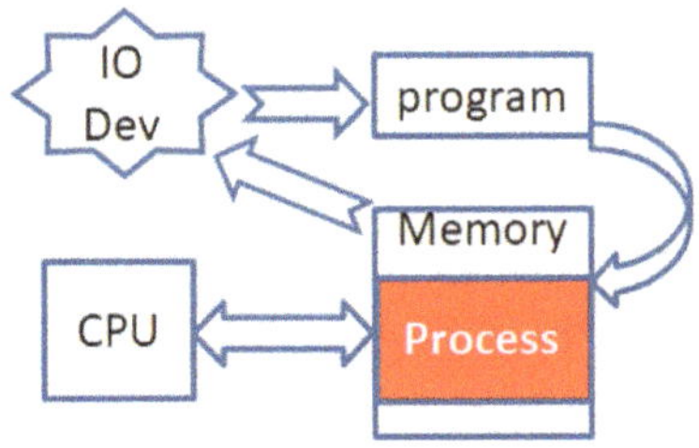

FIGURE 13.1 Architecture of Basic Computer System

This fetch-decode-execute procedure is performed in a cycle, and the cycle is repeated until the program is finished. During the process, the program counter (PC), a special register inside the CPU, plays an important role in program control (order of instruction execution). In the fetch step, the CPU must check the value of the PC before fetching. The PC value is the memory address that the CPU will use to access the next instruction to be executed. Before a program is executed, the initial PC value is the starting address of the program in the memory space. Once the CPU finishes the first cycle (fetches and executes the first instruction of the program), by default, the PC value would be increased by a word size automatically. If it is a 64-bit system whose word size is 8 bytes, the PC value would be increased by 8. If it is a 16-bit system, the PC value is increased by 2 after each cycle. In this way, the CPU can eventually access all instructions of a program to finish the execution.

Figure 13.2 is a more detailed example of a 16-bit CPU execution where the PC value is updated after each cycle.

Example 13.1

Before a cycle is performed, the data inside registers R_0 and R_3 are 0005H and 0F00H (in hexadecimal), respectively. Also, the value of the PC register is 0008H. Before the CPU fetches, it uses the PC to get the memory address and then retrieves 2 bytes from the memory addresses 0008H and 0009H, respectively. The decoder translates the fetched 2 bytes 1000 0001 and 0000 0011 into an instruction R3 = R3 + R0, which updates the value in R_3 by adding the value in R_0. The execution step within this cycle updates R_3 with 0F05H (0F00H + 0005H). Thus, the result of this cycle updates the R_3 register. Furthermore, the PC value would be changed automatically to 000AH because it is increased by 2 bytes (0008H+2).

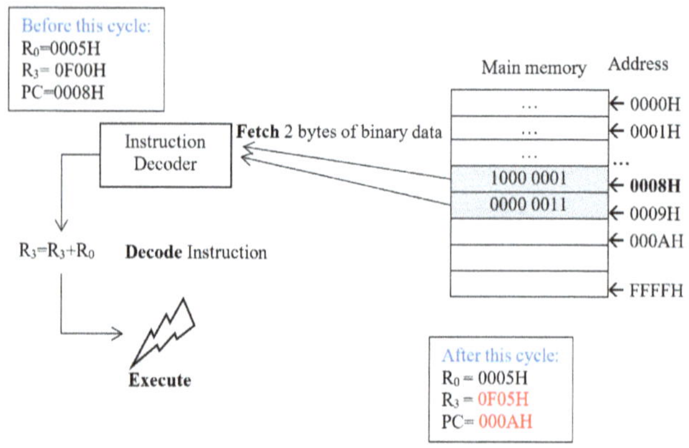

FIGURE 13.2 Example of a 16-bit CPU Cycle

The SimX System and Architecture

We introduce a simplified CPU simulator, the SimX CPU. It is a 16-bit system consisting of an ALU, registers, and a control unit. This SimX architecture comprises 64KB (65536 bytes) memory units with the addresses ranging from 0 (0000H) to 65535 (FFFFH). Figure 13.3 shows the structure of the SimX system.

The SimX CPU is a simulator of a CPU, with basic functions working as a computer, i.e., executing a program by performing cycles according to the Von Neumann architecture, using the fetch-decode-execution cycle (see Figure 13.3).

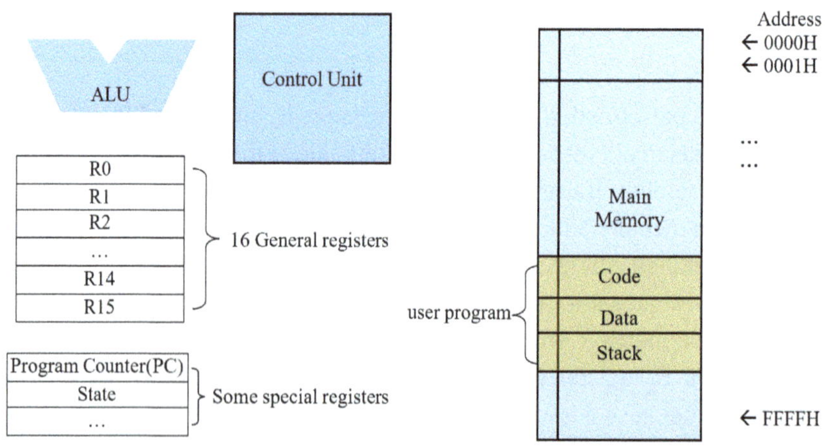

FIGURE 13.3 Diagram of the SimX System

The word size is 16 bits (2 bytes) for the SimX CPU. Thus, nearly all operations operate on 16-bit chunks of data, and all values and addresses are 16 bits in size. Note that 16-bit values are encoded in **big-endian** format, meaning that the most significant byte comes first.

Besides the 16 general-purpose registers (R_0–R_{15}), the CPU has two special 16-bit registers. One is a PC, which stores the address of the *next instruction that will be executed*. The other is called status (*state*), which stores bit flags representing the state of the CPU. The following example illustrates the meaning of the PC value:

Example 13.2

PC Register:	0	1	0	1	0	0	0	1	1	1	0	1	1	0	1	0

Here, the 16-bit binary of the PC value is 01010001 11011001; its hex representation is 51DAH. Therefore, the CPU will fetch and execute the instruction stored in the memory units whose addresses are 51DAH and 51DBH.

For the state register, even though there are 16 bits available, the current setting of the SimX system only utilizes 2 bits as the indicators of the CPU status.

Debug mode ↓

State Register:	0	0	0	0	0	0	0	0	0	0	0	0	0	0	x	x

↑ Condition flag

The least significant bit is the *condition flag*, which represents the status of the last logical test operation. The bit is set TRUE (1) if the condition is true and is FALSE (0) otherwise. The second least significant bit indicates whether the debug mode is on or off. If the debug mode is on (1), then after each instruction is executed, the SimX system prints out the values of all its registers for debugging purposes. In real-world cases, there are more special registers designed in the CPU. Some examples might be bits indicating overflow and carry-out for arithmetic operations. As a simulator, we only showcase these two.

Example 13.3

If the state register value in hexadecimal is 0002H:

State Register:	0	0	0	0	0	0	0	0	0	0	0	0	0	0	1	0

The result of previous performed testing is False, and the CPU is in the debug mode.

Additionally, the SimX CPU uses the last general-purpose register, R_{15}, to store the address (pointer) to the program stack. Note that the terms *pointer* and *address* both refer to a location in memory. Although the address is a number, we can look at it pictorially as pointing to the word in memory. When the stack is empty, its pointer references the bottom of the program space. This register is incremented by two (due to the 16-bit system) when an item is popped off the stack and decremented when an item is pushed on the stack. Figure 13.4 shows example of stack operations and the status of the stack pointer.

Example 13.4

Suppose there is a stack space in the memory with 256 bytes. The stack pointer is an address ranging from 0500H to 05FFH. Originally, the stack is empty (at the left of the figure), and the stack pointer is at the bottom of the stack space; its address is 05FFH. After pushing two 16-bit data items, 0315H and 0410H, into the stack in order, the stack pointer is moved up (decreased) by four steps to the nearest empty space, whose address is 05FBH. After a pop operation, the data item on the top of the stack 0410H (0000 0100 0000 0010 in binary) is removed from the stack, and the stack pointer is moved down (increased) by two steps to the space with the address 05FDH. Note that the value of the stack pointer in the SimX system is stored in register R_{15}.

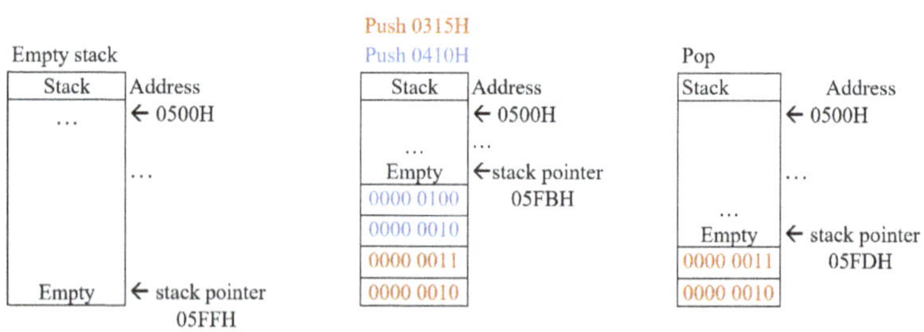

FIGURE 13.4 Example of Stack Operation

The size of the program stack is specified by the code being executed. The program stack is used to store temporary values, arguments to a function, and the return address when a function call is made.

The SimX Instruction Set

In this section, we introduce a simplified RISC-based instruction set for this 16-bit SimX system. Specifically, it has a small number (approximately 34) of instructions, most of which are 2 bytes in size.

As explained in Example 13.4, CPU takes cycles to execute the instructions defined in the program by performing fetch, decode, and execute repeatedly. Every unit in the main memory has an address. The instruction is fetched from the main memory according to the *PC* value, which is the memory address. After each cycle, by default, the *PC* value is increased. Thus, CPU is able to fetch the next instruction until it reaches the end of the program (program is finished). Here, we first introduce the classes of instructions.

Instruction Lists

This instruction set defines 34 instructions that perform arithmetic and logic, data movement, stack manipulation, and control flow. Those instructions can be categorized into four classes: 3 zero-operand instructions, 11 one-operand instructions, 17 two-operand instructions, and 3 extended instructions.

Zero-Operand Instructions

Table 13.1 lists three zero-operand instructions. Their mnemonics are:

- ret: return to the previous address after a function call is finished.
- cld: clear the debug bit (set to 0) in the state register.
- std: set the debug bit in the state register to 1.

TABLE 13.1 Zero-Operand Instructions

MNEMONIC	ENCODING	DESCRIPTION
ret	00000001 *0*	Return from a procedure call
cld	00000010 *0*	Clear the debug bit
std	00000011 *0*	Set the debug bit

Each zero-operand instruction takes 1 word (2 bytes) space. The format of the encoding for a zero-operand instruction is:

2	6	8
00	######	00000000

The two most significant bits of the first byte are 00, and the following 6 bits ###### represent the instruction identifier. All the bits of the second byte are zeros. There is no operand needed.

Example 13.5

If CPU fetches 2 bytes from the memory like the following:

00	000001	00000000

the decoder sees the 00 in the two most significant bits; it knows it is a zero-operand instruction and will further check the following 6 bits. In this case, the decoder will know it is a "return" command because 000001 matches the encoding definition; **ret** instruction would pop out the address from the stack into the PC, which is previously pushed in. Thus, the program would be back to the previous code block.

One-Operand Instructions

Table 13.2 lists 11 one-operand instructions. Note that ***S*** and ***D*** in the "Encoding" column are 4-bit binary, representing registers *rS* and *rD*.

Each one-operand instruction takes 1 word (2 bytes) space. The format of the encoding has two variants:

	2	1	5	4	4
Variant 1:	01	0	#####	RRRR	0000

and

	2	1	5	8
Variant 2:	01	1	#####	vvvvvvvv

The two most significant bits of both variants are 01 (indicating a one-operand instruction), the third significant bit indicates whether the operand is an immediate (a value stored in the instruction itself) or a register, and the following five bits ##### represent the instruction identifier. For variant 1, the third most significant bit is 0,

TABLE 13.2 One-Operand Instructions

MNEMONIC	ENCODING	DESCRIPTION	MEANING
neg *rD*	01000001 *D 0*	Negate register *rD*	*rD* = −*rD*
not *rD*	01000010 *D 0*	Logically negate register *rD*	*rD* = !*rD*
inc *rD*	01001000 *D 0*	Increment *rD*	*rD* = *rD* + 1
dec *rD*	01001001 *D 0*	Decrement *rD*	*rD* = *rD* − 1
push *rS*	01000011 *S 0*	Push register *rS* onto the program stack	*r15* = *r15* − 2 memory[*r15*] = *rS*
pop *rD*	01000100 D *0*	Pop value from stack into register *rD*	*rD* = memory[*r15*] *r15* = *r15* + 2
jmpr *rS*	01000101 *S 0*	Jump to address in register *rS*	*PC* = *rS*
callr *rS*	01000110 *S 0*	Call to address in register *rS*	*r15* = *r15* − 2 memory[*r15*] = *PC* *PC* = *rS*
out *rS*	01000111 *S 0*	Output the character stored in the low order byte from the register *rS* to stdout	output ← *rS*
br *L*	01100001 *L*	Branch relative to label *L* if condition bit is true	if *F* == *0x0001* then *PC* = *currentAdr* + *L*, where F is the Condition Flag in the State Register.
jr *L*	01100010 *L*	Jump relative to label *L*	*PC* = *currrentAdr* + *L*

the operand is a register identifier, and the four most significant bits of the second byte encode the register ID that is ranged from 0000 to 1111 in binary.

Example 13.6

For the given 2-byte data:

2	1	5	4	4
01	0	01001	0101	0000

The decoder would recognize it as a one-operand instruction: **dec** $\mathbf{R_5}$, because

- the two most significant bits are **01**
- the third most significant bit is 0, which is a variant 1 instruction

- According to the SimX ISA (Table 13.2), this instruction is **dec** because the following 5 bits are 01001.
- Since the most significant 4 bits of the second byte is 0101, which is 5 in decimal, register R_5 is used.

dec R_5 is to perform decrement on the value stored in register R_5 by one.

For variant 2 of a one-operand instruction, the third most significant bit is 1; the second byte encodes the immediate value as the operand. It means this instruction uses the second byte as the operand instead of a value stored in a certain register.

Example 13.7

For the given 2-byte instruction:

2	1	5	8
01	1	00010	1100 0000

The decoder would recognize it as a one-operand instruction, **jr** 192, because

- the two most significant bits are 01.
- the third most significant bit is 1, which is a variant 2 instruction.
- according to the design (Table 13.2), this instruction is **jr** because the following 5 bits are 00010.
- since the second byte is 1100 0000, which is 192 in decimal, 192 is used as the operand.

jr 192 adds 192 to the current PC value. Thus, the program could jump forward to a different code block.

Two-Operand Instructions

Table 13.3 lists 17 two-operand instructions. Note that **S** and ***D*** in the "Encoding" column are 4-bit binary, representing registers *rS* and *rD*. **S** represents Source (where an operand value is coming from), and ***D*** represents Destination (where the result will be stored).

TABLE 13.3 Two-Operand Instructions

MNEMONIC	ENCODING	DESCRIPTION	MEANING
add *rS*, *rD*	10000001 *S D*	Add register *rS* to register *rD*	*rD* = *rD* + *rS*
sub *rS*, *rD*	10000010 *S D*	Subtract register *rS* from register *rD*	*rD* = *rD* – *rS*
mul *rS*, *rD*	10000011 *S D*	Multiply register *rD* by register *rS*	*rD* = *rD* * *rS*
div *rS*, *rD*	10000100 *S D*	Divide register *rD* by register *rS*	*rD* = *rD*/*rS*
and *rS*, *rD*	10000101 *S D*	And register *rS* with register *rD*	*rD* = *rD* & *rS*
or *rS*, *rD*	10000110 *S D*	Or register *rS* with register *rD*	$rD = rD \mid rS$
xor *rS*, *rD*	10000111 *S D*	Xor register *rS* with register *rD*	*rD* = *rD* ^ *rS*
shr *rS*, *rD*	10001000 *S D*	Shift right register *rD* by register *rS*	*rD* = *rD* >> *rS*
shl *rS*, *rD*	10001001 *S D*	Shift left register *rD* by register *rS*	*rD* = *rD* << *rS*
test *rS1*, *rS2*	10001010 *S1 S2*	Set condition flag to true if and only if *rS1* ! = *rS2*	if *rS1* ! = *rS2* then *F* = *True* else *F* = *False*
cmp *rS1*, *rS2*	10001011 *S1 S2*	Set condition flag to true if and only if *rS1* < *rS2*	if *rS1* < *rS2* then *F* = *True* else *F* = *False*
equ *rS1*, *rS2*	10001100 *S1 S2*	Set condition flag to true if and only if *rS1* equals *rS2*	if *rS1* == *rS2* then *F* = *True* else *F* = *Flase*
mov *rS*, *rD*	10001101 *S D*	Copy register *rS* to register *rD*	*rD* = *rS*
load *rS*, *rD*	10001110 *S D*	Load word into register *rD* from memory pointed to by register *rS*	*rD* = memory[*rS*]
stor *rS*, *rD*	10001111 *S D*	Store word from register *rS* to memory at address in register *rD*	memory[*rD*] = *rS*
loadb *rS*, *rD*	10010000 *S D*	Load byte into register *rD* from memory pointed to by register *rS*	*rD* = (byte) memory[*rS*]
storb *rS*, *rD*	10010001 *S D*	Store byte from register *rS* to memory at address in register *rD*	(byte)memory [*rD*] = *rS*

A two-operand instruction also takes 1 word (2 bytes) space. The format of the encoding is:

2	6	4	4
10	######	SSSS	DDDD

where the two most significant bits are 10 (indicating a two-operand instruction), and the following six bits ###### represent the opcode. The second byte encodes the two registers as the locations of the operands in two 4-bit chunks. Each 4-bit chunk identifies one of the 16 registers (0–15).

Example 13.8

Given the following 2-byte data:

2	6	4	4
10	000001	0001	0011

The decoder is able to recognize it as a two-operand instruction: **add** R_1, R_3, because

- the two most significant bits are 10
- the following 6 bits, 000001, represent the opcode **add**
- the first 4-bit chunk of the second byte is 0001, which means the first operand is stored in register R_1
- the second 4-bit chunk of the second byte is 0011 in binary, which is 3 in decimal. It means the second operand is stored in register R_3.

add R_1, R_3 adds the value in the first register R_1, and the value in the second register R_3. The value in R_3 register will be updated: $R_3 = R_1 + R_3$

Extended Instructions

The extended instructions have 2 words (4 bytes). The label or value is encoded as a single word following the word containing the instruction. If the instruction is an *extended* instruction, the PC needs to be incremented by an additional 2 bytes during the instruction's execution. Table 13.4 lists three extended instructions. Note that **D** in the "Encoding" column of the **loadi** mnemonic is 4-bit binary representing register *rD*, while *a* is a 16-bit value.

TABLE 13.4 Extended Instructions

MNEMONIC	ENCODING	DESCRIPTION	FUNCTION
jmp *L*	11000001 *0*	Absolute jump to label *L*	*PC* = *L*
call *L*	11000010 *0*	Absolute call to label *L*	*r15* = *r15* – 2 memory[*r15*] = *PC* *PC* = *L*
loadi *a*, *rD*	11100001 *D 0*	Load immediate value or address *a* into register *rD*	*rD* = *a*

The format of the encoding also has two variants:

	2	1	5	8	16 bits
Variant 1:	11	0	#####	00000000	L

and

	2	1	5	4	4	16 bits
Variant 2:	11	1	#####	RRRR	0000	L

The two most significant bits of both variants are 11 (indicating an extended instruction). The third significant bit indicates whether a second register operand is used or not, and the following 5 bits ##### encode the instruction opcode.

For variant 1, the third most significant bit is 0. The variant 1 of the extended instruction only uses the one-word (16-bit) immediate operand that follows the first word.

Example 13.9

The following 4 bytes show the variant 1 of an extended instruction:

2	1	5	8	16 bits
11	0	00001	00000000	0011 1100 0001 1110

The two most significant bits are 11, and its instruction ID is (00001). Since the third significant bit is 0, this instruction's operand is stored in the rightmost 16 bits, 0011 1100 0001 1110. From **Table 13.4**, this is a jump instruction; its mnemonic is JMP. The JMP instruction will cause the CPU to go to a different code block, either several steps backward or a few steps ahead, to fetch and execute the instruction. In programming, JMP helps to implement a *loop* or *if-else* statement by changing the program control flow. The operand 0011 1100 0001 1110 is the address in this example and will be copied to the PC register. Thus, instead of accessing the next instruction in a sequential order, the CPU can fetch and execute the code stored in the specified address. In this way, a "Jump" is implemented.

For variant 2, the third most significant bit is 1. This is a two-operand extended instruction. The four most significant bits of the second byte encode the register ID (0000 ~ 1111) that is the one operand of the instruction. Another operand is the one-word (16-bit) immediate operand that follows the first word.

Example 13.10

The following 4 bytes are for variant 2 of an extended instruction:

2	1	5	8	16 bits
11	1	00001	10000000	0011 1100 0001 1110

The two most significant bits are 11, and its instruction ID is (00001). Since the third significant bit is 1, the first 4 bits of the second byte $(1000)_2$ are the register R_8 as the one operand. The other operand is stored in the rightmost 16 bits, 0011 1100 0001 1110. From Table 13.4, this is a data load instruction,

loadi 15390, r8 # $(0011\ 1100\ 0001\ 1110)_2 = (15{,}390)_{10}$

This instruction loads 16-bit data (0011 1100 0001 1110), which is 15,390 in decimal, into register R_8.

Assembly Programming

Assembly language uses mnemonics to represent low-level machine instructions. Using the designed RISC-based SimX system mnemonics and a set of syntax rules, we can write SimX system assembly code.

Following are the rules and regulations for this assembly language:

1. # mark is the comment sign. Anything to the right of a # mark, including the #, is considered a comment and ignored by the assembler.
2. Blank lines are ignored.
3. Each line in the assembly file that is not blank must contain a directive, a label, and/or an instruction. If the line contains both, the label must precede the instruction.
4. A label is of the form *string*, where *string* consists of any sequence of letters (A-Za-z), digits (0–9), or underscores (_) as long as the first character is not a digit. A colon (:) must terminate the label. A label represents the corresponding location in the program and may be used to jump to that particular location in the code.
5. An instruction consists of a mnemonic, such as **add**, **loadi**, **push**, etc., followed by 0 or more operands. The operand must be separated from the mnemonic by one or more white-space characters. Multiple operands are separated by a comma. See *lines 02* and *03* in Example 13.11.

Example 13.11

In the following code

```
..   ...
1      loop:          # loop through string
2    loadb r0, r1 # load bye
3    test  r1, r1 # check if not nil byte
4    br    output # if byte is not nil, branch to output
5    jr    done   # else branch to done
6      output:
7    out   r1     # output byte
8    inc   r0     # increment pointer
9    jr    loop   # loop
10     done:
11     ret            # return
```

Both "loop" and "output" in *lines 1* and 6 are the labels, which are followed by a colon (:). The statement "br output" in *line 4* forces the program to jump to "out r1" in *line 7.*

6. The mnemonics are case-sensitive. All mnemonics are in lowercase according to the tables. For instance, the assembler does not recognize **ADD** or **Add**.
7. If an operand is a register, then it must be in the form **r#** where # ranges between 0 and 15, inclusively. E.g., **r7** refers to register 7.
8. If the instruction is immediate, then the argument may either be a label or an integer. Note: Labels are case-sensitive. If a label is specified, no colon should follow the label.
9. Directives instruct the assembler to perform a specific function or behave in a specific manner. All directives begin with a period and are followed by a keyword. There are three directives defined in this SimX assembly language, each of which takes an operand:

   .literal

   .words

   .glob

a. The ***.literal*** directive encodes a null-terminated string or an integer at the present location in the program. If the ***.literal*** is followed by an integer, it encodes a nil-terminated string followed by a 16-bit (1 word) value representing the decimal value.

Example:

mystring:
.literal "Hello World!"
myvalue:
.literal 42

Here, there are two labels preceding each of the encodings so that it is easy to access these literals. That is, the label *mystring* represents the address (relative to the start of the program) where the string "Hello World!" is encoded, and the label *myvalue* represents the address (relative to the start of the program) of the decimal value 42. The usage is demonstrated in the following hello.asc example.

Example 13.12

LINE NO.	ASSEMBLE CODE: HELLO. ASC	COMMENTS
1	`.glob puts`	# external function ***puts*** to be used
2		
3	`.glob main`	# export main
4	`main:`	
5	`loadi hellostr, r0`	# load pointer to the string
6	`push r0`	# push pointer as a parameter for ***puts***
7	`call puts`	# call puts to output string
8	`call puts`	# call puts to output string
9	`call puts`	# call puts to output string
10	`call puts`	# call puts to output string
11	`pop r0`	# remove the parameter in the stack
12	`ret`	# return the main function
13	`hellostr:`	
	`.literal "Hello World!"`	

This assembly program is to print out "Hello World!" on the screen four times. Additionally, "puts" is an external print function defined in another code file (see it in appendix). It is imported or included into this program by the **.glob** directive at line 1. Pay attention to the last two lines of the program: hellostr is a label that is used as the address for the following .literal directive. String "Hello World!" is the content of the address named as the label "hellostr". In *line 5*, the statement *loadi hellostr, r0* puts the address of the "Hello World!" string into register R_0. The push instruction puts the address on the top of the stack. This will allow the puts print function to access the string by popping the stack.

b. The ***.words*** directive sets aside a specified number of words of memory at the present location in the program. E.g.,

myspace:
.words 6

allocates 6 words of memory at the present point in the program. This space is not initialized to any specific value. Just as before, the label preceding the directive represents the address of the first word, relative to the start of the program. This is used to set aside space for the program stack.

c. The ***.glob*** directive exports the specified symbol (label) if it is defined in the file and imports the specified symbol (label) if it is used but not defined in the file. E.g.,

.glob **foo**
.glob **bar**
...
loadi bar, r0
...
foo:
.literal "Hello World!"

This example declares two symbols (labels): **foo** and **bar**. Symbol ***foo*** is declared in this file, so it will be exported (can be accessed) in other files. The latter symbol, ***bar***, is used but not defined. When this file is linked, the linker looks for the symbol (label) definition in other files and makes all references to the symbol refer to where it is defined.

Note: We recommend that you place literals and all space allocations at the end of your program so that they will not interfere with the program itself. If you do place literals in the middle of your program, you will need to ensure that your code jumps around these allocations.

A Sample Assembly Program: arithmetic.asc

We illustrate and explain a simple assembly program: *arithmetic.asc*. This program shows how the data is loaded into registers to perform various arithmetic calculations, including addition, subtraction, multiplication, and several bit operations. Meanwhile, the binary codes of the corresponding mnemonic statements are also presented to the right.

According to the design of the SimX instruction set, each statement has direct mapping with the binary code. The assembler converts the assembly program file into binary code (see Example 13.13).

Example 13.13

TABLE 13.5 Assembly Code and Its Corresponding Binary Code

LINE	ASSEMBLY CODE	EXPLANATION	MACHINE CODE	(HEX)
1	`std`	#set the debug mode on	0000 0011 0000 0000	0300
2	`loadi Stack, r15`	#set register R15 as the stack address.	1110 0001 1111 0000	E1F0
3			0000 0010 0011 0101	0235
4	`loadi 2, r0`	#save value 2 into register R0	1110 0001 0000 0000	E100
5			0000 0000 0000 0010	0002
6	`loadi 3, r1`	#save value 3 into register R1	1110 0001 0001 0000	E110
7			0000 0000 0000 0011	0003
8	`loadi 4, r2`	#save value 4 into register R2	1110 0001 0010 0000	E120
9			0000 0000 0000 0100	0004
10	`loadi 5, r3`	#save value 5 into register R3	1110 0001 0011 0000	E130
11			0000 0000 0000 0101	0005
12	`add r2, r3`	#R3=R2+R3 = 4+5=9	1000 0001 0010 0011	8123
13	`sub r2, r3`	#R3=R3 - R2=9-4=5	1000 0010 0010 0011	8223
14	`mul r2, r1`	#R1=R2*R1 = 4*3=12	1000 0011 0010 0001	8321
15	`div r2, r1`	#R1=R1/R2 = 12/4=3	1000 0100 0010 0001	8421
16	`neg r0`	#R0= - R0 = -2	0100 0001 0000 0000	4100
17	`neg r0`	#R0= - R0 = 2	0100 0001 0000 0000	4100
18	`mov r2, r4`	#R4 = R2 = 4	1000 1101 0010 0100	8D24
19	`loadi 7, r5`	# save value 7 into register R5	1110 0001 0101 0000	E150

LINE	ASSEMBLY CODE	EXPLANATION	MACHINE CODE	(HEX)
20			0000 0000 0000 0111	0007
21	`and r3, r0`	#R0= R3 bit and with R0; R0=4 & 2 = 0	1000 0101 0011 0000	8530
22	`or r2, r1`	#R1=R2 bit or with R1; R1=4 | 3 =7	1000 0110 0010 0001	8621
23	`xor r5, r2`	#R2=R5 xor R2, R2=7 xor 4 =3	1000 0111 0101 0010	8752
24	`not r4`	#Not R4, R4=32,763	0100 0010 0100 0000	4240
25	`not r4`	#not R4 again, R4=4	0100 0010 0100 0000	4240
26	`shl r2, r5`	#shift left r5 by r2, R5=7<<3=56	1000 1001 0010 0101	8925
27	`shr r2, r5`	#shift right R5 by R2, R5=56>>3=7	1000 1000 0010 0101	8825
28	`inc r5`	# R5=7+1=8	0100 1000 0101 0000	4850
29	`dec r5`	#R5=8-1=7	0100 1001 0101 0000	4950
30				
31	`.literal 0`	#put 1byte 0 (halt) as the end of the program	0000 0000 0000 0000	0000
			...[512-byte space]...	...
32	`.words 512`	#set 512 bytes as the stack space		
33	`Stack:`	#it is labeled as the bottom of the stack		

Chapter Summary

A simplified RISC-based instruction set architecture is introduced in this chapter. These instructions can be categorized into four classes:

- *Zero-operand* instructions
  - *1 word (2 bytes)*
  - 3 instructions
  - 2 most significant bits are 00

00	######	...

- *One-operand* instructions
  - *1 word (2 bytes)*
  - 11 instructions
  - 2 most significant bits are 01

01	######	...

- *Two-operand* instructions
  - *1 word (2 bytes)*
  - 17 instructions
  - 2 most significant bits are 10

10	######	...

- *Extended* instructions
  - *2 words (4 bytes)*
  - 3 instructions
  - 2 most significant bits are 11

11	######	...

According to the mnemonics and the assembly programming syntax rules, students can understand some of the provided assembly programs and can also write some simple assembly code.

Chapter 13 Exercises

1. Suppose we have a 16-bit CPU, with 8 regular registers, and their current values are:

R0 = 0 × 0002	R1 = 0 × 0000	R2 = 0 × 1101	R3 = 0 × 0009
R4 = 0 × 0001	R5=0 × 0002	R6 = 0 × 0000	R7 = 0 × 0009

Also, the following cells represent some parts of the memory, where each cell contains one byte, and the addresses are labeled on the right side. E.g., the value in address 1000 is 0101 1010. Here, the address is in hexadecimal format.

...	
0101 1010	←addr: 1000
...	...
1000 0010	←addr: 1008
0100 0101	←addr: 1009
1000 1001	←addr: 100A
0000 0010	←addr: 100B
...	
...	

Based on the von Neumann model, answer the following questions:

a. Current PC register value is 0 × 1008, what is the meaning of this number?
b. What is the data that CPU is going to fetch? Explain your answer.
c. What is the PC value after the current cycle? Explain your answer.

MNEMONIC	ENCODING	DESCRIPTION
...	...	...
add *rS*, *rD*	10000001 ssssdddd	Add register *rS* to register *rD*.
sub *rS*, *rD*	10000010 ssssdddd	Subtract register *rS* from register *rD*.
mul *rS*, *rD*	10000011 ssssdddd	Multiply register *rD* by register *rS*.
div *rS*, *rD*	10000100 ssssdddd	Divide register *rD* by register *rS*.
and *rS*, *rD*	10000101 ssssdddd	And register *rS* with register *rD*.
or *rS*, *rD*	10000110 ssssdddd	Or register *rS* with register *rD*.
xor *rS*, *rD*	10000111 ssssdddd	Xor register *rS* with register *rD*.
shr *rS*, *rD*	10001000 ssssdddd	Shift right register *rD* by register *rS*.
shl *rS*, *rD*	10001001 ssssdddd	Shift left register *rD* by register *rS*.
...	...	

d. What is the data CPU is going to fetch in the next cycle? Explain your answer.
e. What is the CPU operation according to the fetched data in the next CPU cycle (check the mnemonic table to get the decoded result)?
f. What is the execution result of the operation in question e? (List the value of related register.) Explain your answer.
g. Explain the steps of the fetch-decode-execute cycle. Your explanation should associate with the value changes in various registers in question f.

2. Suppose we have a 16-bit CPU with 8 regular registers, and their current values are:

R0 = 0 × 0002	R1 = 0 × 000F	R2 = 0 × 1101	R3 = 0 × 0009
R4 = 0 × 0001	R5 = 0 × 0002	R6 = 0 × 0000	R7 = 0 × 0009

Indicate the register being changed and its new value after each of the following assembly language instructions. Show the results in hexadecimal.

a. neg r0
b. inc r1
c. add r2, r3
d. mov r7, r6
e. xor r5, r4
f. or r5, r7

Appendix: The SimX Utilities

To help compile and link programs and finally generate the machine-executable binary code, a set of utilities (programs) are provided, including an assembler, a linker, and a combination of assembler and linker.

SimX Assembler (simas)

Note: Computer systems can only understand and execute the binary machine codes. To translate an assembly language program into binary code, an assembler is required. An assembler (**simas**) is provided to allow you to write and compile programs for the SimX CPU.

The format of the assembler is:

`simas` <*assembly file*> <*binary file*>

Suppose there is an assembly program called *example.asc*:

LINE NO.	ASSEMBLE CODE: EXAMPLE.ASC	COMMENTS
1	`loadi 72, r3`	# load a decimal number 72 into register R3
2	`loadi 4, r4`	# load a decimal number 4 in to register R4
3	`out r3`	# print 'H' on the screen because its ASCII code is 72
4	`add r4, r3`	# make the number in R3 being 76
5	`out r3`	# print 'L' on the screen because its ASCII code is 76

The program prints two characters, 'HL', on the screen. To compile this simple assembly program, the assembler should be invoked:

`./ simas` *example.asc example.s*

This invocation will cause the assembler to read in the file *example.asc* and generate an output file *example.s*, which can be executed in a simulator of an operation system (explained later), where two characters 'HL' would be printed on the screen.

SimX Linker (simld)

For some programs that need to combine multiple program files or include some existing libraries, a linker tool is required. The linker is responsible for taking the assembled .aso object files and linking them together into one large object file or executable. It is useful when your program is spread across multiple files. The format of the linker is:

`simld` *<target> <start> <object files>*

The following are parameters:

- **target**: the destination object/executable file
- **start**: the object file (.aso) where execution should start
- **object files**: one or more additional object files (.aso) containing code that you wish to be linked together.

Suppose there is an assembly program called *hello.asc*:

LINE NO.	ASSEMBLE CODE: HELLO.ASC	COMMENTS
1	`.glob puts`	# external function ***puts*** to be used
2		
3	`.glob main`	# export main
4	`main:`	
5	`loadi hellostr, r0`	# load pointer to the string
6	`push r0`	# push pointer as a parameter for ***puts***
7	`call puts`	# call puts to output string
8	`pop r0`	# remove the parameter in the stack
9	`ret`	# return the main function
10		
11	`hellostr:`	
12	`.literal "Hello World!"`	

In the program, an external function "puts" is called in line 6 to print out a string "Hello World!"on the screen. Function "puts" is defined in another assembly program, stdio.asc:

ASSEMBLE CODE: STDIO.ASC	COMMENTS
`.glob puts`	
puts:	
`push r0`	# save registers to be used
`push r1`	
`mov r15, r1`	# save stack pointer
`loadi 6, r0`	# add 3 words to access params
`add r0, r1`	
`load r1, r0`	# load param into r0
`loop:`	# loop through string
`loadb r0, r1`	# load byte
`test r1, r1`	# check if not nil byte
`br output`	# if byte is not nil, branch to output
`jr done`	# else branch to done
`output:`	
`out r1`	# output byte
`inc r0`	# increment pointer
`jr loop`	# loop
`done:`	
`loadi 0xa, r1`	# output newline
`out r1`	#
`pop r1`	# restore r1
`pop r0`	# restore r0
`ret`	# return

Meanwhile, a start-up program is used to combine both:

LINE NO.	ASSEMBLE CODE: START.ASC	COMMENTS
1	`.glob main`	#
2		
3	`  loadi Stack, r15`	# setup the stack pointer
4	`  call main`	# call main function defined in hello.asc
5		
6	`.literal 0`	
7		
8	`.words 512`	#initialize the stack space(512 bytes)
9	`Stack:`	
10		

To create an executable binary code of *hello.asc, the* following steps are required:

1. First, use the assembler, simas, to create intermediate object files for hello.asc, stdio.asc, and start.asc (start-up code)

   ```
   ./ simas hello.asc  hello.aso
   ./ simas stdio.asc  stdio.aso
   ./ simas start.asc  start.aso
   ```

   Here, .aso files are the object binary files. The order of generating these object files does not matter.

2. Then, the linker tool is invoked to combine three programs together:

   ```
   ./ simld hello.s  start.aso hello.aso stdio.aso
   ```

which would create an executable (for the SimX system) file called *hello.s* containing *start.aso* (start-up code), *hello.aso* (main function), and *stdio.aso* (output function).

The simcc

If compiling and linking multifile programs are required, the ***simcc*** utility is helpful. It is analogous to **gcc,** but just for our simulation system, SimX CPU. The ***simcc*** utility assumes that the code begins in the main function, just like the *hello.asc* example, and that the SimX object files (*stdio.aso*) are in the same directory as the program files.

You can then use the ***simcc*** utility to assemble and link all needed files. ***simcc*** will perform this process in one step instead of using both *simas* and *simld*. For example, to create the *hello* program, we specify the target object/executable file via the **-o** switch and then specify all required assembly or object files.

```
./simcc -o hello.s hellow.asc stdio.asc
```

Then, **simcc** will figure out what to assemble and what to link. Note that there is no need to specify the *start.asc*, file if *hello.asc* follows the default code structure (define a main function).

The Simulation Environment (simXOS)

The SimX system provides a virtual environment for executing the assembled and compiled SimX binary code. This environment, **simXOS,** simulates a simple computer with a SimX CPU equipped with a RISC-based instruction set introduced in Chapter 12. The **simXOS** allocates memory space, loads a "binary" file, and begins executing it. The simulator system stops after a specified number of cycles or when an illegal operation is performed.

The simXOS takes two parameters on the command line: The first parameter is a nonnegative integer denoting the maximum number of cycles that a program should run for, and the remaining argument is the object/executable file of the program to run. For example, the invocation

```
./simXOS 1000 hello.s
```

Then, **simXOS** will load the binary code *hello.s* and run the program for at most 1000 cycles. If 0 cycles are specified, this is interpreted as an infinite number of cycles. It means the program will run until all instructions are performed.

www.ingramcontent.com/pod-product-compliance
Ingram Content Group UK Ltd.
Pitfield, Milton Keynes, MK11 3LW, UK
UKHW052008041026
15590UKWH00008B/2218

9 781793 561541